Fortschritte der Chemie organischer Naturstoffe

Progress in the Chemistry of Organic Natural Products

60

Founded by L. Zechmeister
Edited by W. Herz, G. W. Kirby, R. E. Moore,
W. Steglich, and Ch. Tamm

Authors:
C. A. A. van Böckl, A.-M. Eklund, M. Petitou,
I. Wahlberg

Springer-Verlag
Wien New York 1992

Prof. W. Herz, Department of Chemistry,
The Florida State University, Tallahassee, Florida, U.S.A.

Prof. G. W. Kirby, Chemistry Department,
The University, Glasgow, Scotland

Prof. R. E. Moore, Department of Chemistry,
University of Hawaii at Manoa, Honolulu, Hawaii, U.S.A.

Prof. Dr. W. Steglich, Institut für Organische Chemie und Biochemie der Universität
Bonn, Bonn, Federal Republic of Germany

Prof. Dr. Ch. Tamm, Institut für Organische Chemie der Universität Basel,
Basel, Switzerland

© 1992 by Springer-Verlag/Wien

Library of Congress Catalog Card Number AC 39-1015

Typesetting: Macmillan India Ltd., Bangalore-25

Printed on acid free paper

With 59 Figure

ISBN-13: 978-3-7091-9227-6   e-ISBN- 978-3-7091-9225-2

DOI: 10.1007/ 978-3-7091-9225-2

Contents

List of Contributors

VAN BOECKL, Dr. C.A.A., Organon Scientific Development Group P.O. Box 20, NL-5340 Oss, The Netherlands.

EKLUND, Dr. A.-M., Reserca AB, Box 17007, S-104 62 Stockholm, Sweden.

PETITOU, Dr. M., Sanofi Research, Rue de Président S. Allende, F-94256 Gentilly, France.

WAHLBERG, Dr. I., Reserca AB, Box 17007, S-104 62 Stockholm, Sweden.

Cyclized Cembranoids of Natural Occurrence

I. WAHLBERG and A.-M. EKLUND, Reserca AB, Stockholm, Sweden

Contents

I. Introduction

The structure of eunicillin (**89**), a carbobicyclic diterpenoid isolated from the Mediterranean gorgonian *Eunicella stricta*, was reported in 1968 (*1*). At that time chlorine-containing diterpenoids had been discovered in the gorgonian *Briareum asbestinum* (*2, 3*), but it was not until 1977 that the structure of the first briaran, briarein A (**211**), was resolved by X-ray analysis (*4*). The first trinervitane diterpenoid (**26**), which was isolated from *Trinervitermes* termites, was reported in 1976 (*5*).

89   211   26

The discovery of these compounds marked the appearance of a large and growing group of diterpenoids which are commonly viewed as being formed from cembrane precursors by secondary carbon-carbon bond closures. It should be emphasized, however, that in the absence of biosynthetic evidence the question of which groups of diterpenoids that should be classified as cyclized cembranoids remains ambiguous. In the present article we have included five tobacco diterpenoids (**1–5**) which. possess prerequisite structural features and cooccur with appropriate cembrane precursors in tobacco. Although structurally reminiscent of cyclized cembranoids, verticillanes, taxanes and cleomeolide are not dealt with, since these diterpenoids of plant origin are not believed to arise *via* preformed cembranoids (*6–11*).

The bi-, tri- and tetracyclic secotrinervitanoids, trinervitanoids and kempanoids, which are present in the defensive secretions of soldiers of higher termites, are most likely formed via cyclization of cembrane precursors (*12*). It is also generally agreed that diterpenoids such as briarans, cladiellins and asbestinins are correctly characterized as cyclized cembranoids (*13–15*). These compounds, which show a large structural diversity, are constituents of marine invertebrates.

We have previously reviewed the cembranoids of natural occurrence (*16*). Our intention in the present review is to follow a similar outline and to give a comprehensive compilation of the naturally occurring cyclized cembranoids as defined above, that have appeared in the literature through December 1991. Biogenetic relationships are also discussed. It should be added that a review on cyclized cembranoids was published by RALDUGIN and SHEVTSOV in 1987 (*17*), but their selection of diterpene classes differs from ours.

The cyclized cembranoids are frequently heavily substituted and contain several asymmetric centers. It has therefore been necessary to use X-ray diffraction methods for elucidation of the stereostructures of many compounds. References to these X-ray studies are given in Tables 1–3.

Like their presumed cembrane precursors, many cyclized cembranoids and particularly those of marine origin exhibit important biological and pharmacological effects, this being one of the reasons for the great interest in their structures and chemistry. Available references are given in Table 3.

Very few cyclized cembranoids have been prepared synthetically. KATO *et al.* (*18–20*) have completed the syntheses of two secotrinervitanoids (**7**, **9**) and DAUBEN *et al.* (*21*) have recently published the total synthesis of ($\pm$)-kempene-2 (**59**). Another diterpenoid of insect origin, longipenol (**64**), has been the target of a synthetic study (*22*) as has the tobacco basmanoid 3 (see Tables 1 and 2) (*23, 24*).

A. Nomenclature and Structural Representation

The present article includes more than two hundred compounds belonging to as many as sixteen different diterpene classes. The nomenclature systems and principles employed in the literature vary among the classes and also within certain classes. In the case of the cyclized cembranoids of insect origin, the secotrinervitane, trinervitane and kempane nomenclatures are generally accepted (*25*). Nomenclatures for the diterpene classes found in tobacco have also been established (*26–28*). In addition to existing trivial names, we have therefore introduced a semi-systematic nomenclature based on skeletal type for the tobacco and insect constituents listed in Tables 1 and 2, respectively. The *R*- and *S*-system has been adopted to describe the configuration of each compound.

The situation is different for the cyclized cembranoids of marine origin. This is illustrated by the fact that some authors (*14, 29, 30*) employ a briaran or briarein nomenclature based on the oxygen-containing skeleton **A**, while a briarane nomenclature based on the carbon skeleton **B** has been adopted by others (*31*). Similarly, both the asbestinin (**C**) and asbestinane (**D**) nomenclatures are found in the literature (*14, 32, 33*). Because of this lack of consensus, we have only included trivial names

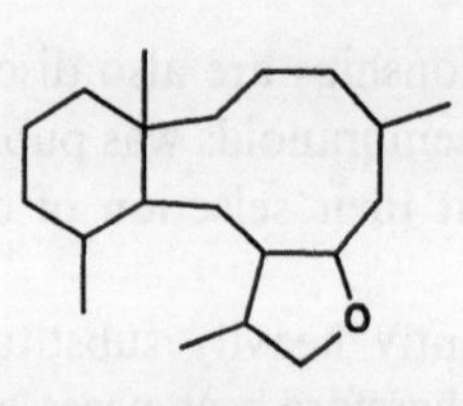

A

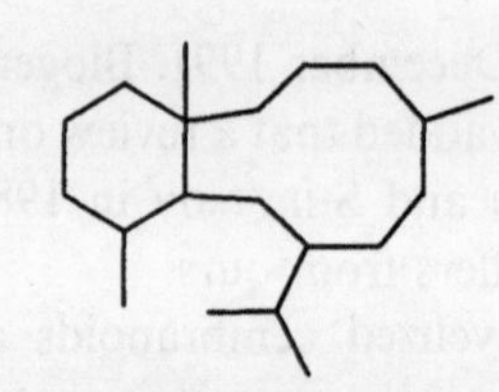

B

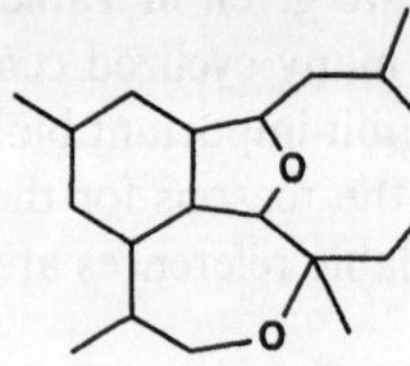

C

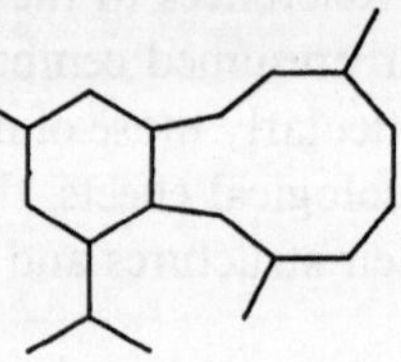

D

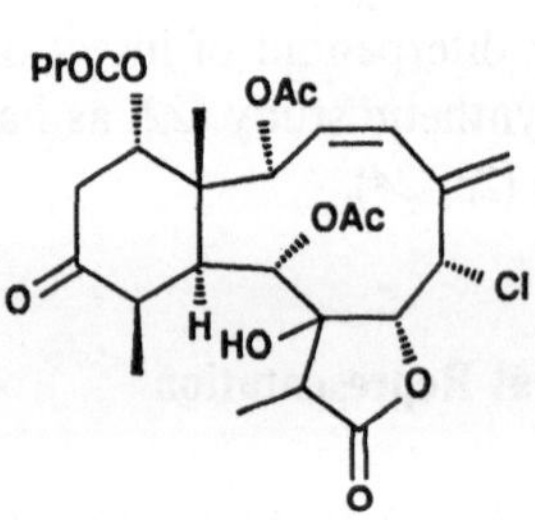

191a

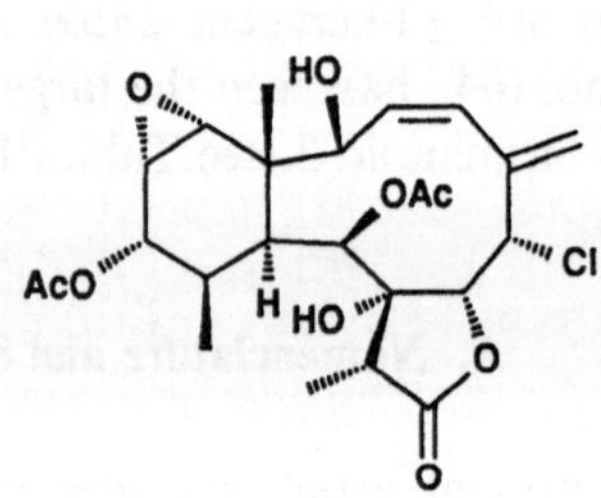

192a

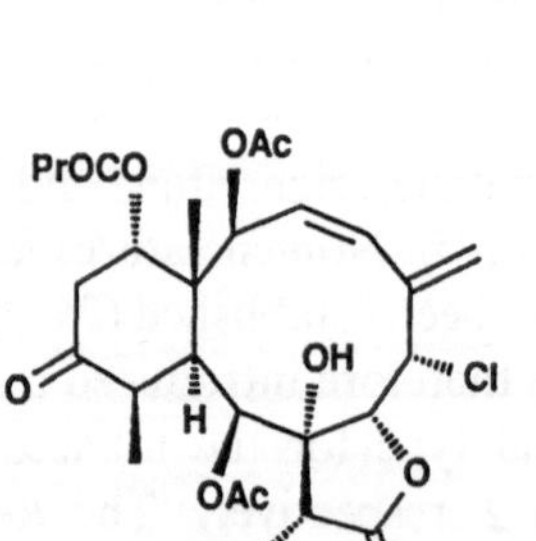

191

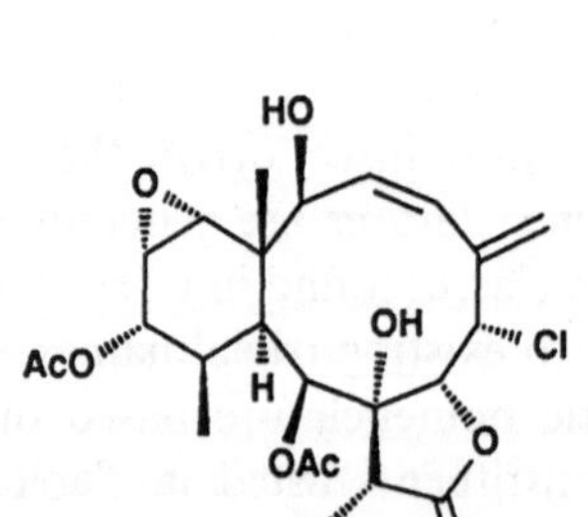

192

and configurational *R*- and *S*-descriptors in Table 3. For unnamed compounds, however, a semi-systematic nomenclature based on the skeletal types defined in Section IV has been adopted.

The graphical representation of macrocyclic rings is not always straightforward. This may lead to confusion as is illustrated for ptilosarcone and brianthein X. Both compounds have (2*S*,9*S*)-configurations and are represented as **191a** and **192a**, respectively, in the original articles (*34, 35*). In the interest of clarity we have therefore chosen to employ the style shown in structures **191** and **192** for diterpenoids of the briaran class.

II. Cyclized Cembranoids from Tobacco

Five diterpenoids (**1–5**), which are classified as cyclized cembranoids and listed in Table 1, have hitherto been isolated from tobacco (*26–28, 36, 37*). Of these, **1** and **2** possess a carbobicyclic capnosane skeleton (**E**), **3** and **4** a carbotricyclic basmane skeleton (**F**) and **5** a carbotricyclic virgane skeleton (**G**). The names assigned to these skeletal frameworks are generic: capnosane comes from κaπvos, the Greek word for tobacco, basmane from basma, a tobacco leaf class and virgane from Virginia tobacco. While compounds having capnosane type skeletons have also been found in marine organisms, **3** and **4** are the only basmanoids and **5** the only virganoid so far encountered in nature.

E   F   G

Compounds **1–5** possess structural features consistent with the view that they are formed in tobacco by intramolecular cyclization reactions taking place in appropriate cembranic precursors. The latter commonly covered in the gorgonian *Briareum asbestinum* (2, 3), but it was not until substituents at C-4 and C-6 (*16*). The biogenesis of the capnosanoid **2** can therefore be envisioned to proceed via a transannular reaction connecting C-3 with C-7 in a parent cembranoid. The 11,12 double bond, although of Z-geometry in **2**, is retained and oxygenation is, not unexpectedly, found not only at C-4 and C-6 but also at C-2, C-8 and C-10 (*26*). The capnosanoid **1**, which does not contain the 11,12 double bond, may be derived from a cembrane precursor such as **229** (Scheme 1). Bond closure across C-3 and C-7 and concurrent ketal formation give **1** (*36*).

The basmanoids **3** and **4** are suggested to arise via cyclization reactions involving all three double bonds in the cembranoid precursor

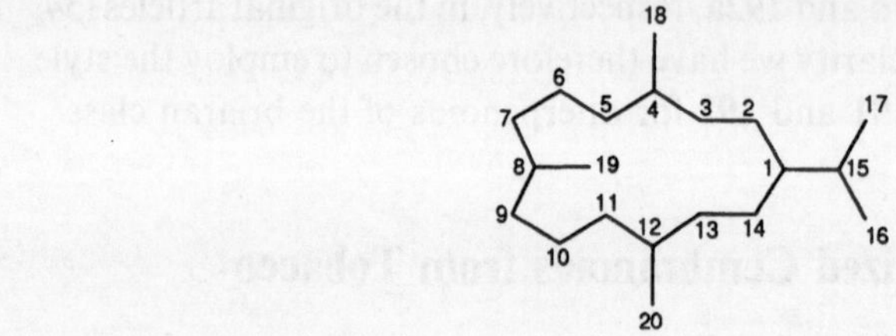

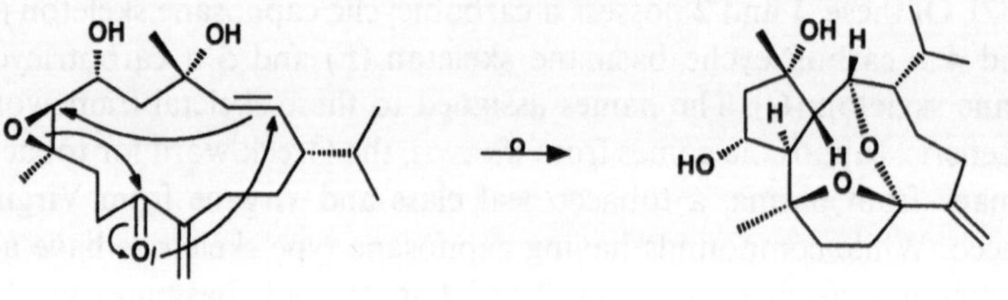

Scheme 1. Proposed biogenesis of the capnosanoid **1**

and resulting in the formation of new carbon-carbon bonds between C-2 and C-12 and between C-7 and C-11 (*27, 37*).

As proposed in Scheme 2, the biosynthetic route to the virganoid **5** is initiated by an allylic rearrangement reaction converting the 4,6-diol **230**,

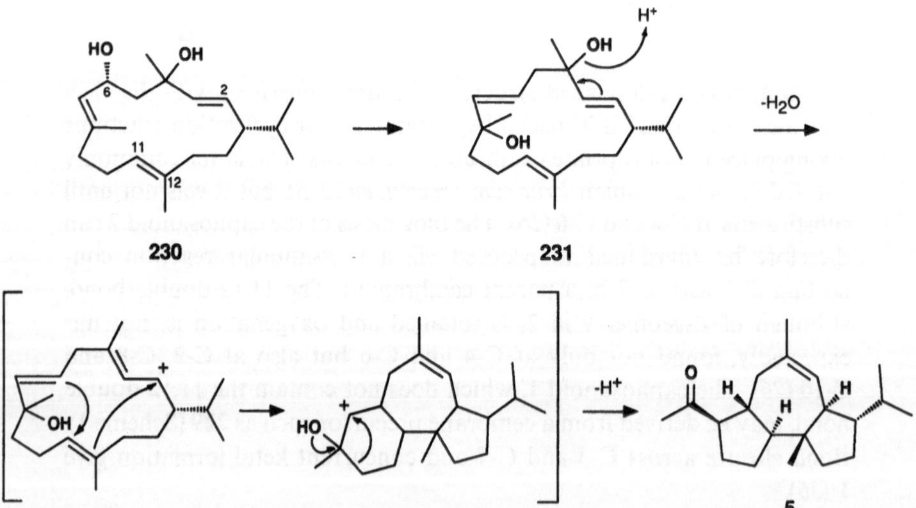

Scheme 2. Proposed biogenesis of the virganoid **5**

Table 1. *Cyclized Cembranoids from Tobacco*

Structure no.	Compound Name	Ref.	X-Ray Study
3,7-Cyclized Cembranoids; Capnosanoids			
1	(1S*,2R*,3S*,4S*,6R*,7R*,8R*,11R*)-2,11:8,11-Diepoxycapnos-12(20)-ene-4,6-diol	(36)	X-ray analysis of **1** (relative configuration) (36)
2	(1S*,3R*,4S*,6R*,7R*,8R*,11Z)-4,6,8-Trihydroxycapnos-11-ene-2,10-dione	(26)	X-ray analysis of the benzoate of **2** (relative configuration) (26)
2,12:7,11-Cyclized Cembranoids; Basmanoids			
3	(1S*,2S*,4Z,7R*,8S*,11R*,12R*)-7,8-Epoxy-4-basmen-6-one	(27)	X-ray analysis of **3** (relative configuration) (27)

Table 1 *(continued)*

Structure no.	Compound Name	Ref.	X-Ray Study
4	$(1R^*,2S^*,3R^*,4S^*,7S^*,8S^*,11R^*,12S^*)$-1,3-Epoxy-4,8-dihydroxybasman-6-one	(37)	X-ray analysis of **4** (relative configuration) (37)
5	*2,12:6,11-Cyclized Cembranoid; Virganoid* $(1S^*,2S^*,3Z,6R^*,7R^*,10S^*,11R^*)$-Virg-3-en-18-one	(28)	X-ray analysis of the semicarbazone of **5** (relative configuration) (28)

which is a major tobacco cembranoid (*38*), to a 4,8-diol **231**. The latter, which is also present in tobacco (*39*), undergoes dehydration. Subsequent acid-induced cyclization connecting C-2 with C-12 and C-6 with C-11 and a pinacol-type of rearrangement give the virganoid **5** (*28*).

III. Cyclized Cembranoids from Insects

Higher termites (family: Termitidae; subfamily: Nasutitermitinae) are well-established sources of cyclized cembranoids. More than fifty compounds (**6–64**) have been reported to-date, all being unique to termites (see Table 2). These compounds fall into five groups: the carbobicyclic secotrinervitanes (**H**) (*40*), the carbotricyclic trinervitanes (**I**) (*41, 42*) and the carbotetracyclic kempanes (**K**) (*42, 43*), rippertane (**L**) (*42, 44*) and longipane (**M**) (*45*). The trinervitanoids form the largest group comprising more than forty compounds (**11–53**), the other groups showing considerably less structural diversity.

These diterpenoids are present together with monoterpene hydrocarbons in the defensive secretions ejected from frontal glands by nasute termite soldiers. While the monoterpenes serve both as alarm pheromones for members of the same species and as toxic agents against predators, the role of the diterpenoids is less well understood. Available experimental data show that they are less toxic than the monoterpenes and that they may act as a solute which retards the evaporation of the monoterpenes (*46–49*).

The biogenetic origin of the cyclized cembranoids has been studied by injection of uniformly labelled [^{14}C]-D-glucose, and of [1-^{14}C]- and [2-^{14}C]-sodium acetate, [2-^{14}C]-(R,S)-mevalonolactone and [3-^{14}C]-(R)-mevalonolactone into the abdomens of soldiers of *Nasutitermes octopilis*. All substances were incorporated into the terpenoids present in the defense secretions, although at different levels. The highest efficiency was observed for [3-^{14}C]-(R)-mevalonolactone. From the results obtained, PRESTWICH *et al.* (*12*) conclude that the diterpenoids are not of dietary origin but are synthesized *de novo* in the termite soldiers.

As proposed in Scheme 3 (*12, 44, 45, 47, 50, 51*), (R)-cembrene A (**232**), which has been isolated from *Nasutitermes exitiosus* (*52*), *Cubitermes glebae* (*53*) and *C. umbratus* (*54*) and *Trinervitermes bettonianus* (*55*), is a plausible precursor. This is converted to the corresponding (3S,4S)-epoxide **233** which undergoes an acid-induced reaction to form a 4,16 carbon-carbon bond. Loss of a proton gives the secotrinervitanoids **6** and **234**. The latter is converted into the kempane type of compounds (e.g. **59**) by reactions involving the generation of 11,15 and 7,16 bonds and into trinervitanoids such as **235** by reactions connecting C-7 with C-16. The monool **235** may serve as an intermediate in the biogenesis of the spiro-fused longipenol **64** and the methyl-shifted rippertenol **63**.

Support for the validity of the initial sequence in this biogenitic route has been provided by biomimetic synthesis. This has involved treatment of the (1S,2R,3S,4S)-3,4-epoxycembra-7,11,15-trien-2-yl acetate **236** with BF$_3$·OEt$_2$ in diethyl ether. The product obtained was converted to the secotrinervita-7,11,15(17)-triene-2,3-diol **7** by reduction using LAH (Scheme 4) (*19, 20*).

Reviews dealing with various aspect of cyclized cembranoids of insect origin have previously been published by PRESTWICH (*46, 48, 56–58*), BAKER and HERBERT (*59*) and GOH *et al.* (*60*).

A. Nasutitermitinae

1. Bulbitermes Species

Three kempanoids (**55, 56, 59**) have been isolated from the defense secretions of soldiers of the Malaysian *Bulbitermes singaporensis* (*61*).

2. Cortaritermes Species

Soldiers of the South American *Cortaritermes silvestri* have been shown to produce three trinervitanoids (**31, 33, 34**), of which all are substituted by *O*-acetyl or *O*-propanoyl groups at C-3, C-9 and C-13 (*62*).

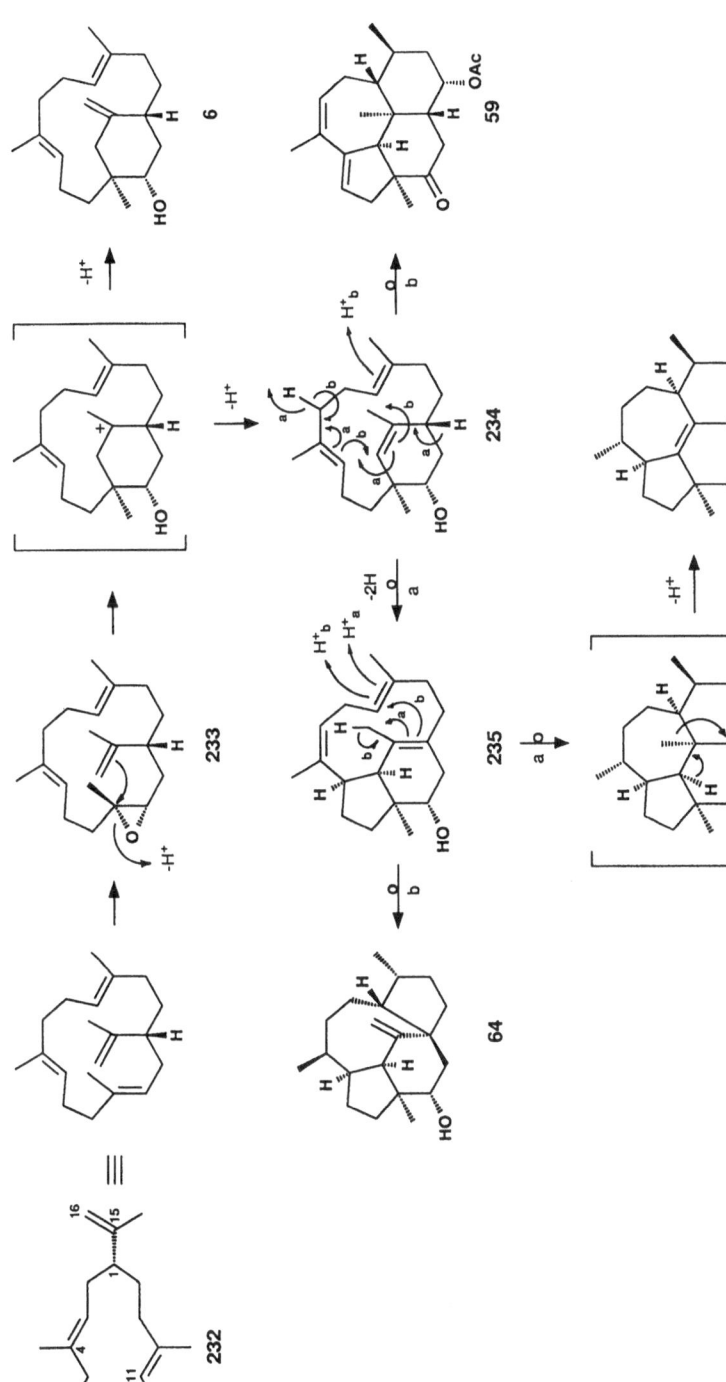

Scheme 3. Proposed biogenesis of the secotrinervitanoid **6**, the kempanoid **59**, rippertenol (**63**) and longipenol (**64**)

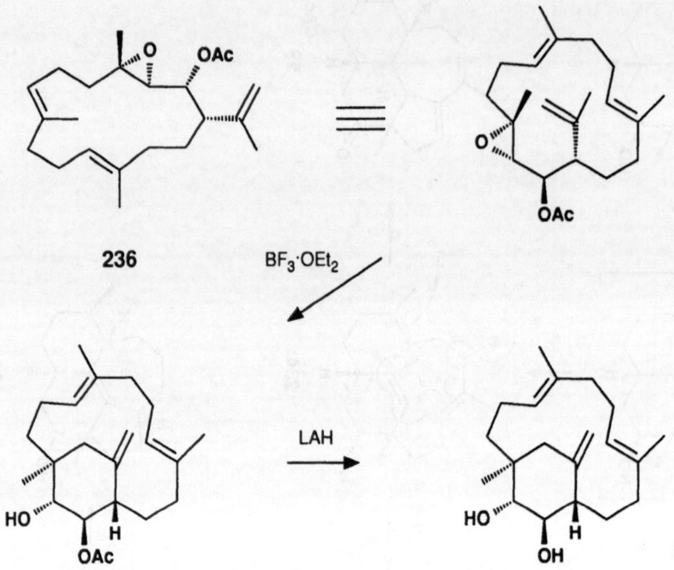

236

BF$_3$·OEt$_2$

LAH

7

Scheme 4. Conversion of the cembranoid **236** into the secotrinervitanoid **7**

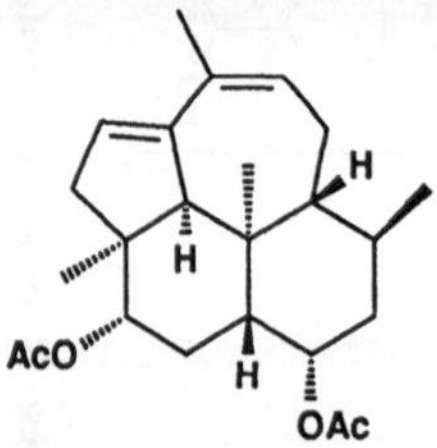

55

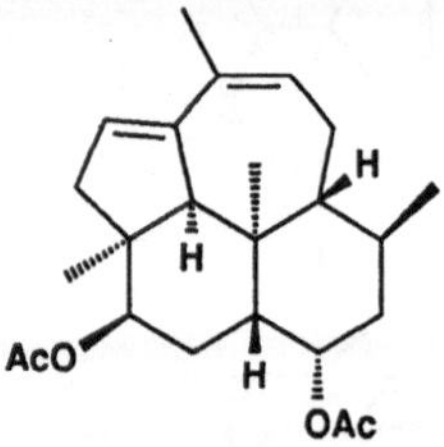

56

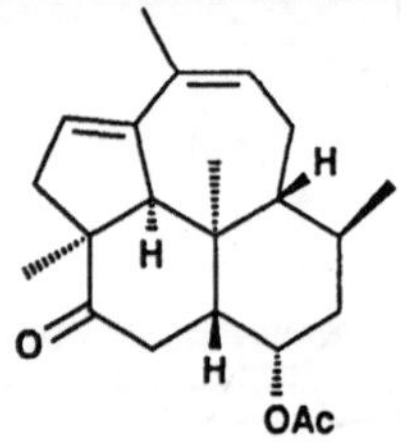

59

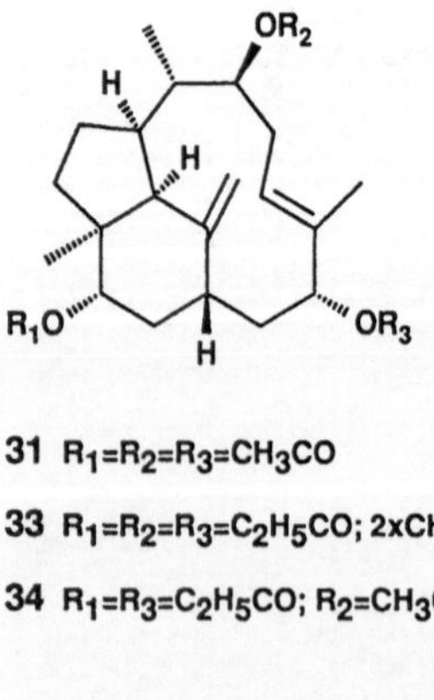

31 R$_1$=R$_2$=R$_3$=CH$_3$CO

33 R$_1$=R$_2$=R$_3$=C$_2$H$_5$CO; 2xCH$_3$CO

34 R$_1$=R$_3$=C$_2$H$_5$CO; R$_2$=CH$_3$CO

3. *Grallatotermes Species*

Trinervitanoids (**11, 12, 14, 21, 24, 26**) and rippertenol (**63**) have been reported as constituents of the defense secretions of soldiers of the East African species *Grallatotermes africanus*. The (2*R*,3*R*)-diol **14** and the diol acetate **26** are the major components. It has been pointed out that, with the exception of **63**, the distribution of diterpenoids in the secretion is similar to that in secretions from *Trinervitermes gratiosus* and *Nasutitermes infuscatus* (*63*).

11   12   14

21   24   63

4. *Hospitalitermes Species*

Four species of the genus *Hospitalitermes* have been examined, these being the Malaysian *H. umbrinus*, *H. hospitalis*, *H. flaviventris* and *H. bicolor* (*64–66*). All elaborate trinervitanoids (e.g. **11, 12, 14, 19, 21, 24, 26, 28, 29, 42, 52, 53**) and rippertenol (**63**). While the diol **14** is the major diterpenoid in the secretions from *H. umbrinus*, *H. hospitalis* and *H. flaviventris*, the secretion from *H. bicolor* contains the diol **21** and the diacetoxymonool **28** as the main components. Noteworthy is the presence of the highly oxygenated trinervitanoids **52** and **53** and, in the case of **53** also a C-17 methylated derivative, in the *Hospitalitermes* genus.

28 R=H

29 R=CH₃CO

42

52 R₁=R₂=C₂H₅CO; R₃=H

53 R₁=C₂H₅CO; R₂=CH₃CO; R₃=CH₃

The intraspecific variation of the chemistry of the defense secretions has been studied for *H. umbrinus* (*65*). It has been shown that the colonies are separable into two groups. Soldiers of one group produce trinerviten-ediols such as **14**, while those of the other group synthesize compounds **52** and **53** which have considerably higher molecular weight (*66*). Although the two groups show no morphological differences, it is noteworthy that soldiers of the former group seem to be more aggressive than soldiers of the latter group.

5. Longipeditermes Species

Longipeditermes longipes, a long-legged, primitive and free-ranging nasute termite in the Malaysian rain forests, is the only known species of this genus. The defense secretions show a large structural diversity with respect to the content of diterpenoids. Four of the five known skeletal types have been encountered, namely secotrinervitanoids (**6, 7**), trinervitanoids (**12, 14, 19, 21, 24, 26, 28, 29**), rippertenol (**63**) and longipenol (**64**). Of these, the secotrinervitatriene-2,3-diol **7** and longipenol (**64**) are unique to *L. longipes* (*45, 47*).

The chemical variability among allopatric and sympatric colonies of this species is also remarkable. It has been reported that while soldiers of one colony may produce predominantly bicyclic secotrinervitanoids and longipenol (**64**), the secretions of soldiers of a nearby colony may contain

6

7

64

trinervitanoids and rippertenol (**63**). The degree of oxygenation of the diterpenoids is also variable among colonies (*47*).

6. *Nasutitermes Species*

The genus *Nasutitermes* includes a large number of species and has an extensive geographical distribution. Several species have been examined for their content of terpenoids, e.g. *N. colombicus* (*67*), *N. corniger* (*67*), *N. costalis* (*43, 68–70*), *N. ephratae* (*44, 67, 69, 71, 72*), *N. gagei* (*73*), *N. gracilirostris* (*74*), *N. havilandi* (*51*), *N. hubbardii* (*49*), *N. infuscatus* (*63, 69*), *N. kemneri* (*62*), *N. kempae* (*69, 75*), *N. lujae* (*76*), *N. luzonicus* (*46*), *N. nigriceps* (*67, 77*), *N. octopilis* (*12, 69, 78*), *N. princeps* (*40, 42*), *N. rippertii* (*44, 69, 78*) and several unnamed *Nasutitermes* species (*42, 74, 80*). In all, more than forty compounds representing the secotrinervitane, trinervitane, kempane and rippertane classes have been discovered.

The defense secretions of some species, e.g. *N. colombicus*, *N. corniger*, *N. ephratae*, *N. nigriceps*, *N. gagei*, and *N. kemneri* are characterized by the presence of mono- and dioxygenated trinervitanoids such as the alcohols **11**, **12**, **14**, **16** and **17** (*62, 67, 73*). By contrast tri- and tetraoxygenated trinervitanoids, mainly esters such as **32**, **34–36**, **43**, **44**, **49–51**, are the major diterpenoids in the secretions from other species such as *N. havilandi*, *N. gracilirostris* and unnamed *Nasutitermes* species from New Guinea (*42, 51, 74*). Soldiers of *N. rippertii* produce a range of mono- to trioxygenated trinervitanoids as well as rippertenol (**63**) (*44, 69, 79*), while *N. princeps* synthesizes mono- to tetraoxygenated trinervitanoids and a fair amount of the secotrinervitanoid **9** (*40, 42*). A secotrinervitanoid (**6**) is also found in the West African *N. lujae* as are the trinervitanoids **12**, **16**, **21** and **23**. Of the latter, the unusual 3,8-epoxide **23** is likely to arise by an intramolecular electrophilic addition of the 3S-hydroxy group on the exocyclic methylene group at C-8 in the diol **16** (*76*).

Kempanoids (**54**, **55**, **57–61**) cooccur with trinervitanoids in secretions from soldiers of *N. kempae*, *N. octopilis*, *N. luzonicus*, *N. infuscatus*, *N. costalis* and unnamed South American *Nasutitermes* species and are the major diterpenoids in the former three species (*43, 46, 62, 69, 70, 75, 78*). Rojofuran (**62**), an unstable furan-containing kempanoid, has been detected among grassland South-American *Nasutitermes* species (*81*).

7. *Subulitermes Species*

Trinervitadiene monools and diols (**11**, **12**, **14**, **16**, **21**) have been discovered in termites of the genus *Subulitermes*, which were collected in Guayana (*73*).

8. Trinervitermes Species

Several species of the genus *Trinervitermes* have been studied (*5, 41, 63, 77, 82–86*). The results show that trinervitanoids, mainly alcohols, are present in the defense secretions (**11–14, 21, 24, 26, 28, 37, 40**). Compositional differences with respect to diterpene content exist between major and minor soldiers of *T. gratiosus*, *T. bettonianus*, *T. geminatus* and *T. togoensis*. In addition, the secretions from the minor soldiers are chemically more complex than those from the major soldiers. In the case of *T. trinervius* and *T. oeconomus*, however, both types of soldiers produce similar secretions (*86*).

13   37   40

9. Velocitermes Species

The Peruvian *Velocitermes velox* is a producer of the trinervitadien-3-ol **11** and the diols **16** and **17**, of which **16** is predominant (*87*).

IV. Cyclized Cembranoids from Marine Invertebrates

Extensive chemical studies have disclosed that marine invertebrates (phylum: Coelenterata: subclass: Octocorallia) form the predominant source of cyclized cembranoids. Up to now such compounds have been encountered in sea fans and sea whips of the order Gorgonacae, in soft corals of the orders Alcyonacae and Stolonifera and in sea pens of the order Pennatulacae. In all more than 160 compounds (**65–228**) belonging to nine diterpene classes have been identified. These are listed in Table 3. The biogenetic relationship between these classes and the cembrane class is indicated in Scheme 5.

Only one 2,6-cyclized cembranoid (**N**) (**65**) has been detected to-date (*89*). The cladiellins (**O**) (*90*) (**66–92**) and sarcodictyins (**P**) (*91, 92*) (**93–98**) form larger groups. They arise by bond closure between C-2 and C-11 and have the same carbon skeleton. A 2,9-epoxy bridge (cladiellin numbering, *93*) is, however, present in compounds of the cladiellin type, while the sarcodictyins have a hemiacetal bridge across C-4 and C-7. These differences in chemical structure are reflected in properties and reactivities, this being the reason for the distinction of the two groups (*91, 92*).

The capnosanoids (**E**) (*26*), which in principle are derivable from a cembranic precursor via a transannular reaction connecting C-3 with C-7, are a small group consisting of four compounds (**110–113**). By contrast, a wide array of briarans (**114–225**), i.e. 3,8-cyclized cembranoids (**A**) (*14, 30*), has been encountered in nature. The majority of these are γ-lactones and many contain a chlorine substituent at C-6 (briaran numbering). They are often highly oxygenated and incorporate ester moieties.

Table 2. Cyclized Cembranoids from Insects

Structure no.	Compound Name	Source	X-Ray Study
4,16-Cyclized Cembranoids; Secotrinervitanoids			
6	(1R*,3S*,4S*,7E,11E)-7,16-Secotrinervita-7,11,15(17)-trien-3-ol	Longipeditermes longipes Nasutitermes lujae	(47) (76)
7	(1S*,2R*,3R*,4S*,7E,11E)-7,16-Secotrinervita-7,11,15(17)-triene-2,3-diol	Longipeditermes longipes	(45, 47)

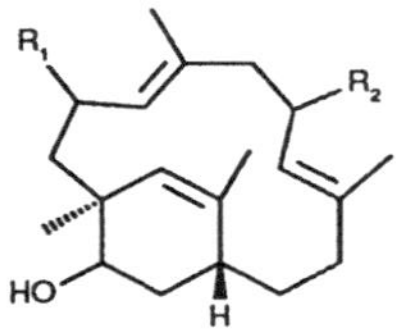

(1R*,4S*,7E,11E)-Acetoxy-7,16-secotrinervita-7,11,15(16)-trien-3-ol

R₁ or R₂ = H or OAc

8

Nasutitermes costalis

(43)

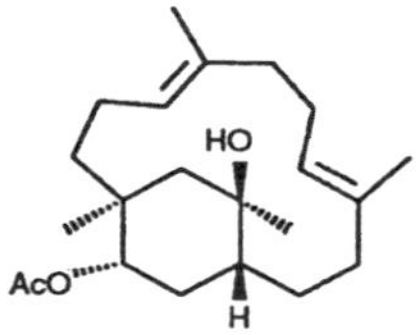

(1R*,3S*,4S*,7E,11E,15R*)-3-Acetoxy-7,16-secotrinervita-7,11-dien-15-ol

9

Nasutitermes princeps

(40, 42)

X-ray analysis of **9** (relative configuration) (40)

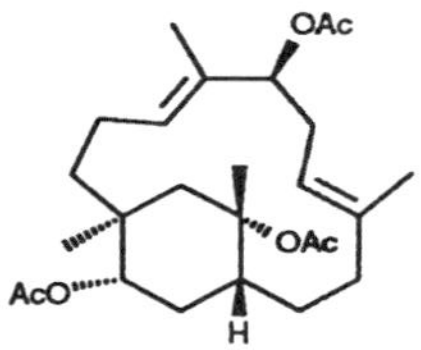

(1R*,3S*,4S*,7E,9S*,11E,15S*)-7,16-Secotrinervita-7,11-diene-3,9,15-triyl triacetate

10

Constrictotermes cyphergaster

(88)

X-ray analysis of the triol derived from **10** (relative configuration) (88)

Table 2 (continued)

Structure no.	Compound Name	Source	X-Ray Study
4.16:7,16-Cyclized Cembranoids; Trinervitanoids			
	(3S,4S,7R,12S,16S)-Trinervita-1(15),8(19)-dien-3-ol	*Grallatotermes africanus*	(63)
		Hospitalitermes flaviventris	(65)
		Hospitalitermes hospitalis	(65)
		Hospitalitermes umbrinus	(64, 65)
		Nasutitermes columbicus	(67)
		Nasutitermes corniger	(67)
		Nasutitermes costalis	(43)
		Nasutitermes ephratae	(67, 69, 71, 72)
		Nasutitermes gagei	(73)
		Nasutitermes hubbardii	(49)
		Nasutitermes infuscatus	(63, 69)
		Nasutitermes kemneri	(62)
		Nasutitermes kempae	(69)
		Nasutitermes nigriceps	(67, 158)
		Nasutitermes octopilis	(69)
		Nasutitermes princeps	(42)
		Nasutitermes rippertii	(79)
		Nasutitermes species n.D	(62)
		Nasutitermes species B	(42)
		Subulitermes baileyi	(73)
		Subulitermes parvellus	(73)
		Trinervitermes bettonianus	(41, 85)
		Trinervitermes geminatus	(86)
		Trinervitermes gratiosus	(41, 63, 82, 84)
		Trinervitermes oeconomus	(86)
		Trinervitermes togoensis	(86)
		Trinervitermes trinervius	(86)
		Velocitermes velox	(87)

11

(3S,4S,7R,8Z,12S,16S)-Trinervita-1(15),8-dien-3-ol — **12**

Grallatotermes africanus	(63)
Hospitalitermes bicolor	(65)
Hospitalitermes flaviventris	(65)
Hospitalitermes hospitalis	(65)
Hospitalitermes umbrinus	(64, 65)
Longipeditermes longipes	(47)
Nasutitermes gagei	(73)
Nasutitermes infuscatus	(63, 69)
Nasutitermes lujae	(76)
Nasutitermes nigriceps	(77, 158)
Nasutitermes octopilis	(69)
Subulitermes parvellus	(73)
Trinervitermes gratiosus	(63, 82)

(3S*,4S*,7S*,8R*,11E,15S*,16S*)-Trinervita-1,11-dien-3-ol — **13**

Trinervitermes oeconomus	(83, 86)
Trinervitermes trinervius	(83, 86)

X-ray analysis of an 11,12-epoxy-3-acetate of **13** (relative configuration) (83)

(2R,3R,4S,7R,12S,16S)-Trinervita-1(15),8(19)-diene-2,3-diol — **14**

Grallatotermes africanus	(63)
Hospitalitermes bicolor	(65)
Hospitalitermes flaviventris	(65)
Hospitalitermes hospitalis	(65)
Hospitalitermes umbrinus	(64, 65)
Longipeditermes longipes	(47)
Nasutitermes columbicus	(67)
Nasutitermes corniger	(67)
Nasutitermes ephratae	(67, 69, 71, 72)

Table 2 *(continued)*

Structure no.	Compound	Name	Source	X-Ray Study
			Nasutitermes gagei	(73)
			Nasutitermes infuscatus	(63, 69)
			Nasutitermes kempae	(69)
			Nasutitermes nigriceps	(67, 158)
			Nasutitermes octopilis	(69)
			Nasutitermes princeps	(42)
			Nasutitermes rippertii	(79)
			Nasutitermes species A	(42)
			Nasutitermes species B	(42)
			Subulitermes oculatissimus	(73)
			Trinervitermes bettonianus	(41, 85)
			Trinervitermes gratiosus	(41, 63, 82, 84)
			Trinervitermes geminatus	(86)
			Trinervitermes oeconomus	(86)
			Trinervitermes togoensis	(86)
			Trinervitermes trinervius	(86)
		(2R*,3S*,4S*,7R*,12S*,16S*)-Trinervita-1(15),8(19)-diene-2,3-diol	*Nasutitermes rippertii*	(72)

15

(2S,3S,4S,7R,12S,16S)-Trinervita-
1(15),8(19)-diene-2,3-diol

Nasutitermes columbicus	(67)
Nasutitermes corniger	(67)
Nasutitermes costalis	(43, 68)
Nasutitermes ephratae	(67, 69, 71, 72)
Nasutitermes gagei	(73)
Nasutitermes hubbardii	(49)
Nasutitermes kemneri	(62)
Nasutitermes lujae	(76)
Nasutitermes nigriceps	(67, 158)
Subulitermes oculatissimus	(73)
Subulitermes parvellus	(73)
Velocitermes velox	(87)

16

(2S,3R,4S,7R,12S,16S)-Trinervita-
1(15),8(19)-diene-2,3-diol

Nasutitermes columbicus	(67)
Nasutitermes corniger	(67)
Nasutitermes costalis	(43, 68)
Nasutitermes ephratae	(67, 71, 72)
Nasutitermes kemneri	(62)
Nasutitermes nigriceps	(67, 158)
Velocitermes velox	(87)

17

Table 2 (continued)

Structure no.	Compound Name	Source	X-Ray Study
18	(2R*,3R*,4S*,7R*,12S*,16S*)-3-Acetoxytrinervita-1(15),8(19)-dien-2-ol	*Nasutitermes rippertii*	(79)
19	(2R*,3R*,4S*,7R*,12S*,16S*)-2-Acetoxytrinervita-1(15),8(19)-dien-3-ol	*Hospitalitermes bicolor* *Hospitalitermes flaviventris* *Hospitalitermes hospitalis* *Hospitalitermes umbrinus* *Longipeditermes longipes* *Nasutitermes rippertii*	(65) (65) (65) (64, 65) (47) (69)
20	(2R*,3R*,4S*,7R*,12S*,16S*)-Trinervita-1(15),8(19)-diene-2,3-diyl diacetate	*Nasutitermes rippertii*	(79)

References, pp. 132–141

(2R*,3R*,4S*,7R*,8Z,12S*,16S*)-
Trinervita-1(15),8-diene-
2,3-diol

Species	Ref.
Grallatotermes africanus	(63)
Hospitalitermes bicolor	(65)
Hospitalitermes flaviventris	(65)
Hospitalitermes hospitalis	(65)
Hospitalitermes umbrinus	(64, 65)
Longipeditermes longipes	(47)
Nasutitermes infuscatus	(63, 69)
Nasutitermes kempae	(69)
Nasutitermes lujae	(76)
Nasutitermes nigriceps	(158)
Subulitermes baileyi	(73)
Subulitermes oculatissimus	(73)
Subulitermes parvellus	(73)
Trinervitermes bettonianus	(41, 85)
Trinervitermes geminatus	(86)
Trinervitermes gratiosus	(41, 63, 82, 84)
Trinervitermes oeconomus	(86)
Trinervitermes togoensis	(86)
Trinervitermes trinervius	(86)

21

(2S*,3S*,4S*,7R*,8Z,12S*,16S*)-
Trinervita-1(15),8-diene-2,3-diol

Species	Ref.
Nasutitermes kemneri	(62)

22

Table 2 (continued)

Structure no.	Compound Name	Source	X-Ray Study
23	(2S*,3S*,4S*,7R*,8S*,12S*,16S*)-3,8-Epoxytrinervit-1(15)-en-2-ol	Nasutitermes lujae (76)	X-ray analysis of 23 (relative configuration) (76)
24	(3R,4S,7R,12S,16S)-3-Hydroxytrinervita-1(15),8(19)-dien-2-one	Grallatotermes africanus (63)	
		Hospitalitermes bicolor (65)	
		Hospitalitermes flaviventris (65)	
		Hospitalitermes hospitalis (65)	
		Hospitalitermes umbrinus (64, 65)	
		Longipeditermes longipes (47)	
		Nasutitermes columbicus (67)	
		Nasutitermes corniger (67)	
		Nasutitermes costalis (69)	
		Nasutitermes ephratae (67, 69, 71, 72)	
		Nasutitermes infuscatus (63, 69)	
		Nasutitermes kempae (69)	
		Nasutitermes nigriceps (67)	
		Nasutitermes octopilis (69)	
		Nasutitermes princeps (42)	
		Nasutitermes rippertii (69)	

(3S*,4S*,7R*,8Z,12S*,16S*)-3-Hydroxytrinervita-1(15),8-dien-2-one

Nasutitermes species B	(42)
Trinervitermes bettonianus	(85)
Trinervitermes geminatus	(86)
Trinervitermes gratiosus	(63)
Trinervitermes oeconomus	(86)
Trinervitermes togoensis	(86)
Trinervitermes trinervius	(86)
Nasutitermes ephratae	(71, 72)

(2R,3R,4S,7R,9R,12R,16S)-9-Acetoxytrinervita-1(15),8(19)-diene-2,3-diol

Grallatotermes africanus	(63)
Hospitalitermes bicolor	(65)
Hospitalitermes flaviventris	(65)
Hospitalitermes hospitalis	(65)
Hospitalitermes umbrinus	(64, 65)
Longipeditermes longipes	(45, 47)
Nasutitermes infuscatus	(63, 69)
Nasutitermes kempae	(69)
Nasutitermes princeps	(42)
Nasutitermes species A	(42)
Nasutitermes species B	(42)
Trinervitermes bettonianus	(41, 85)
Trinervitermes geminatus	(86)
Trinervitermes gratiosus	(5, 63, 82, 84)
Trinervitermes oeconomus	(86)
Trinervitermes togoensis	(86)
Trinervitermes trinervius	(86)

X-ray analysis of **26** (relative configuration)
(5) CD measurement (absolute configuration) (41)

Table 2 (continued)

Structure no.	Compound Name	Source	X-Ray Study
27	(2R*,3R*,4S*,7R*,9S*,12R*,16S*)-9-Acetoxytrinervita-1(15),8(19)-diene-2,3-diol	Nasutitermes species PNG (D,G)	(74)
28	(2R*,3R*,4S*,7R*,9R*,12R*,16S*)-2.3-Diacetoxytrinervita-1(15),8(19)-dien-9-ol	Hospitalitermes bicolor Hospitalitermes flaviventris Hospitalitermes hospitalis Hospitalitermes umbrinus Longipeditermes longipes Trinervitermes gratiosus	(65) (65) (65) (64, 65) (47) (41, 82, 84)

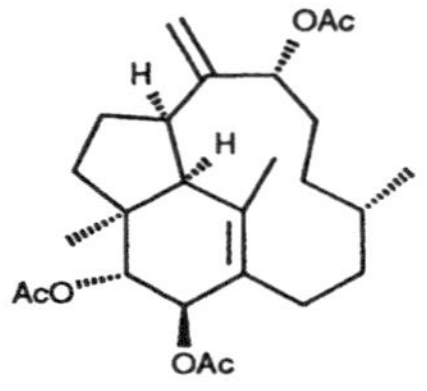

(2R*,3R*,4S*,7R*,9R*,12R*,16S*)-
Trinervita-1(15),8(19)-diene-
2,3,9-triyl triacetate

Hospitalitermes flaviventris (65)
Hospitalitermes umbrinus (64, 65)
Longipeditermes longipes (47)
Nasutitermes hubbardii (49)
Nasutitermes princeps (42)
Nasutitermes rippertii (79)
Nasutitermes species A (42)

29

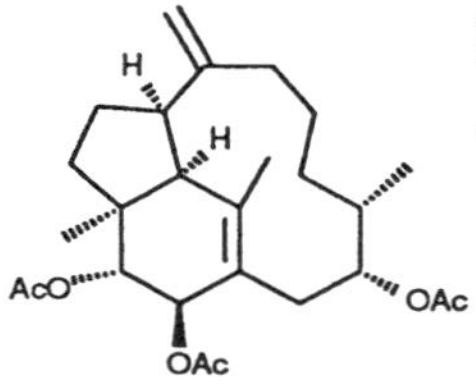

(2R*,3R*,4S*,7R*,12S*,13R*,16S*)-
Trinervita-1(15),8(19)-
diene-2,3,13-triyl triacetate

Nasutitermes nigriceps (158)
Nasutitermes rippertii (79)

30

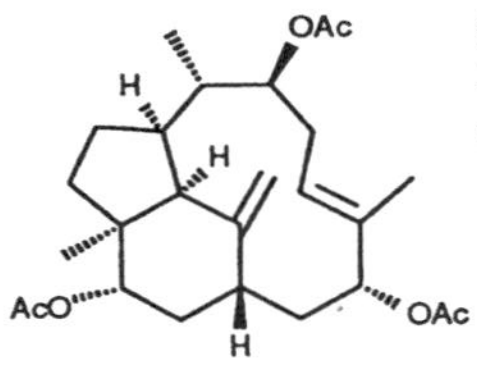

(1R*,3S*,4S*,7R*,8S*,9S*,11E,13R*,
16S*)-Trinervita-11,15(17)-
diene-3,9,13-triyl triacetate

Cortaritermes silvestri (62)

31

Table 2 (continued)

Structure no.	Compound Name	Source	X-Ray Study
32	(1R*,3S*,4S*,7R*,8S*,9S*,11E,13R*, 16S*)-9-Acetoxy-3-propion-oxytrinervita-11,15(17)-dien-13-ol	*Nasutitermes* species C	(42)
33 R=C₂H₅CO; 2×CH₃CO	(1R*,3S*,4S*,7R*,8S*,9S*,11E,13R*, 16S*)-Diacetoxytrinervita-11,15(17)-dienyl propionate	*Cortaritermes silvestri*	(62)
34	(1R*,3S*,4S*,7R*,8S*,9S*,11E,13R*, 16S*)-9-Acetoxytrinervita-11,15(17)-diene-3,13-diyl dipropionate	*Cortaritermes silvestri* *Nasutitermes havilandi* *Nasutitermes* species C	(62) (51) (42)

(1R*,3S*,4S*,7R*,8S*,9S*,11E,13R*,16S*)-3,9-Dipropionoxytrinervita-11,15(17)-dien-13-ol

35

Nasutitermes species C

(42)

(1R*,3S*,4S*,7R*,8S*,9S*,11E,13R*,16S*)-Trinervita-11,15(17)-diene-3,9,13-triyl tripropionate

36

Nasutitermes havilandi
Nasutitermes species C

(51)
(42)

X-ray analysis of the triol derived from **36** (relative configuration) (51)

(2R*,3R*,4S*,7R*,12S*,16S*)-17-Acetoxytrinervita-1(15),8(19)-diene-2,3-diol

37

Trinervitermes bettonianus

(41, 85)

Table 2 (continued)

Structure no.	Compound Name	Source	X-Ray Study
38	(2R*,3R*,4S*,7R*,12S*,16S*)-20-Acetoxytrinervita-1(15),8(19)-diene-2,3-diol	*Nasutitermes* species PNG (F)	(80)
39	(3R*,4S*,7R*,9R*,12R*,16S*)-3,9-Dihydroxytrinervita-1(15),8(19)-dien-2-one	*Nasutitermes princeps*	(42)
40	(3R,4S,7R,9R,12R,16S)-9-Acetoxy-3-hydroxytrinervita-1(15),8(19)-dien-2-one	*Nasutitermes nigriceps* *Nasutitermes princeps* *Trinervitermes geminatus* *Trinervitermes togoensis*	(158) (42) (86) (86)

(3R*,4S*,7R*,12S*,13R*,16S*)-3,13-Dihydroxytrinervita-1(15),8(19)-dien-2-one

Nasutitermes hubbardii (49)

41

(2R*,3R*,4S*,7R*,12S*,16S*)-2,3-Diacetoxytrinervita-1(15),8(19)-dien-13-one

Hospitalitermes bicolor (65)
Hospitalitermes flaviventris (65)
Hospitalitermes hospitalis (65)
Hospitalitermes umbrinus (64, 65)
Nasutitermes rippertii (79)

42

(2R*,3R*,4S*,7R*,9R*,12R*,16S*)-Trinervita-1(15),8(19)-diene-2,3,9,11-tetrayl tetraacetate

Nasutitermes gracilirostris (74)

43

Table 2 *(continued)*

Structure no.	Compound Name	Source	X-Ray Study
44	(2R*,3R*,4S*,7R*,9R*,12R*,16S*)-Trinervita-1(15),8(19)-diene-2,3,9,11-tetrayl tetraacetate	*Nasutitermes gracilirostris*	(74)
45	(2R*,3R*,4S*,7R*,9R*,12S*,13R*,16S*)-Trinervita-1(15),8(19)-diene-2,3,9,13-tetraol	*Nasutitermes princeps*	(42)
46	(2R*,3R*,4S*,7R*,9R*,12S*,13R*16S*)-Trinervita-1(15),8(19)-diene-2,3,9,13-tetrayl tetraacetate	*Nasutitermes princeps* *Nasutitermes* species A *Nasutitermes* species PNG (F)	(42) (42) (74)

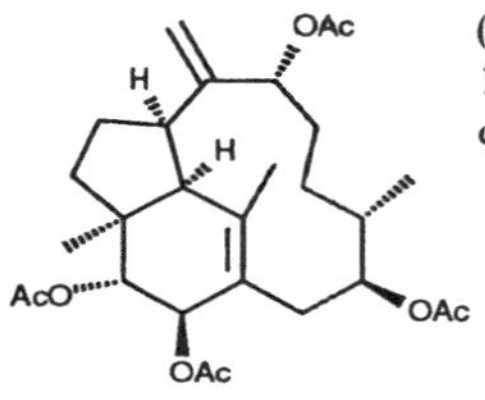

Nasutitermes species PNG (F) (74)

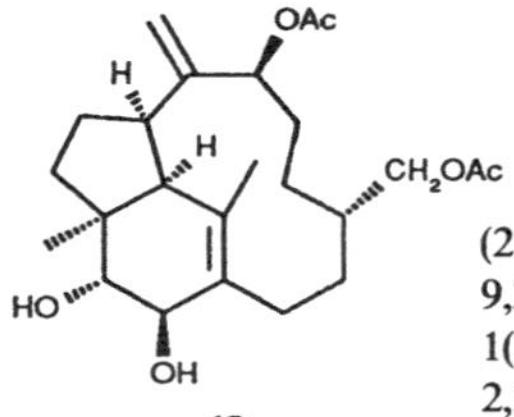

(2R*,3R*,4S*,7R*,9R*,12S*,13S*,
16S*)-Trinervita-1(15),8(19)-
diene-2,3,9,13-tetrayl tetraacetate

47

Nasutitermes species PNG (F) (80)

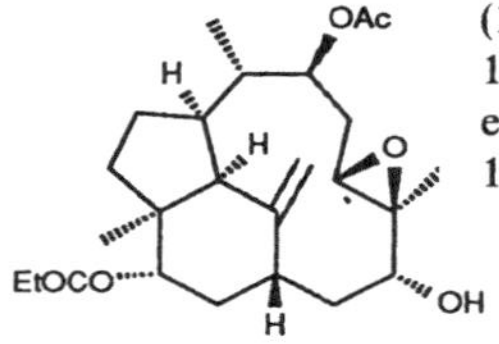

(2R*,3R*,4S*,7R*,9S*,12R*,16S*)-
9,20-Diacetoxytrinervita-
1(15),8(19)-diene-
2,3-diol

48

Nasutitermes species C (42)

(1S*,3S*,4S*,7R*,8S*,9S*,11S*,12S*,
13R*,16S*)-9-Acetoxy-11,12-
epoxy-3-propionoxytrinervita-
15(17)-en-13-ol

49

Table 2 (*continued*)

Structure no.	Compound Name	Source	X-Ray Study
50	(1S*,3S*,4S*,7R*,8S*,9S*,11S*,12S*,13R*,16S*)-9-Acetoxy-11,12-epoxytrinervit-15(17)-ene-3,13-diyl dipropionate	*Nasutitermes* species C	(42)
51	(1S*,3S*,4S*,7R*,8S*,9S*,11S*,12S*,13R*,16S*)-11,12-Epoxytrinervit-15(17)-ene-3,9,13-triyl tripropionate	*Nasutitermes* species C	(42)
52	(2R*,3R*,4S*,7S*,8S*,9R*,12S*,14S*,16S*)-8,19-Epoxy-2,3,9,14,17-pentapropionoxytrinervit-1(15)-en-7-ol	*Hospitalitermes umbrinus*	(66)

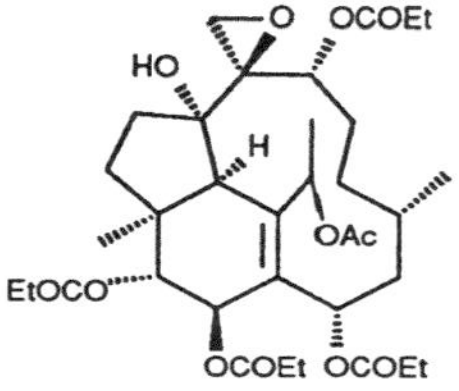

53

(2R*,3R*,4S*,7S*,8S*,9R*,12S*,14S*, 16S*)-17-Acetoxy-8,19-epoxy-17-methyl-2,3,9,14-tetrapropionoxytrinervit-1(15)-en-7-ol

Hospitalitermes umbrinus

(66)

4,16:7,16:11,15-Cyclized Cembranoids; Kempanoids

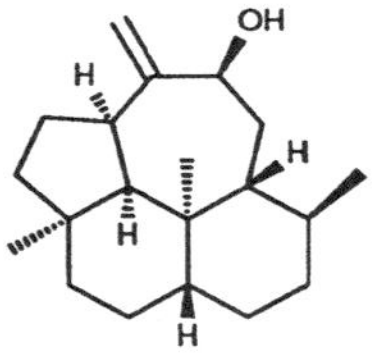

54

(1R*,4S*,7R*,9S*,11R*,12S*,15R*, 16R*)-Kemp-8(19)-en-9-ol

Nasutitermes species n.D

(62)

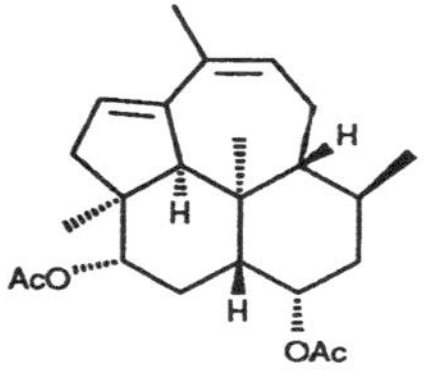

55

Kempene-1
(1R,3S,4S,8Z,11R,12S,14S,15R, 16R)-Kempa-6,8-diene-3,14-diyl diacetate

Bulbitermes singaporensis
Nasutitermes infuscatus
Nasutitermes kempae
Nasutitermes luzonicus

(61)
(69)
(69, 75)
(46)

Table 2 (continued)

Structure no.	Compound Name	Source	X-Ray Study
56	(1R*,3R*,4S*,8Z,11R*,12S*, 14S*,15R*,16R*)-Kempa-6,8-dien-3,14-diyl diacetate	Bulbitermes singaporensis	(61)
57	(1R,3R,4S,7S,8Z,11R,12S,15R,16S)-3-Hydroxykemp-8-en-6-one	Nasutitermes octopilis	(69, 78) X-ray analysis of the p-bromobenzoate of **57** (absolute configuration) (78)
58	(1R*,4S*,8Z,11R*,12S*,14S*,15R*, 16R*)-14-Hydroxykempa-6,8-dien-3-one	Nasutitermes luzonicus	(46)

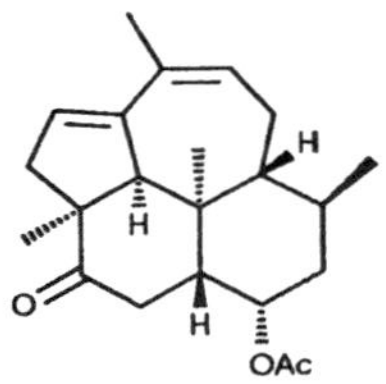

Kempene-2
(1R,4S,8Z,11R,12S,14S,15R,
16R)-14-Acetoxykempa-6,8-dien-
3-one

Bulbitermes singaporensis
Nasutitermes kempae
Nasutitermes luzonicus

(61)
(69, 75)
(46)

X-ray analysis of 59
(relative configuration)
(75) CD measurement
(absolute configuration)
(75)

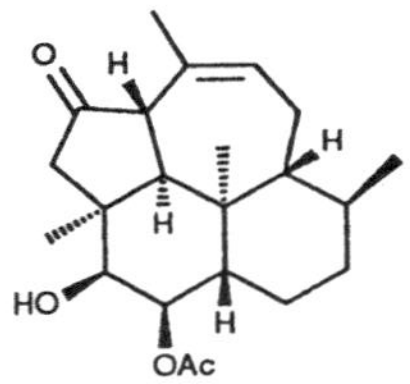

59

(1S*,2R*,3S*,4S*,7S*,8Z,11R*,
12S*,15R*,16S*)-2-Acetoxy-3-
hydroxykemp-8-en-6-one

Nasutitermes octopilis

(69, 78)

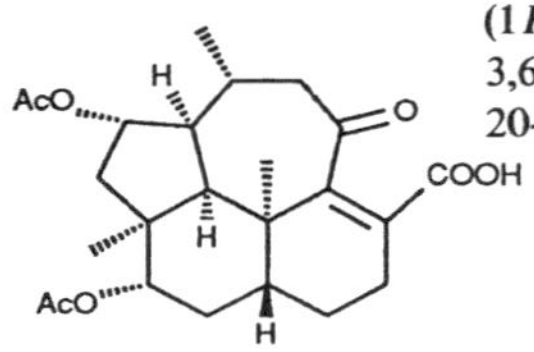

60

(1R*,3S*,4S*,6S*,7S*,8R*,15S*,16S*)-
3,6-Diacetoxy-10-oxokemp-11-en-
20-oic acid

Nasutitermes costalis

(43, 70)

X-ray analysis of the
methyl ester of 61
(relative
configuration) (70)

61

Table 2 (continued)

Structure no.	Compound Name	Source	X-Ray Study
62	Rojofuran (1R*,3R*,4S*,11R*,12S*,15R*,16S*)-6,19-Epoxykempa-6,8(19)-dien-3-yl acetate	Nasutitermes species (81)	
63	4,16:7,16:11,15-Cyclized Cembranoid; Rippertanoid (1S,3S,4S,7S,8R,11S,12S)-Rippert-15-en-3-ol	Grallatotermes africanus (63) Hospitalitermes bicolor (65) Hospitalitermes flaviventris (65) Hospitalitermes hospitalis (65) Hospitalitermes umbrinus (64, 65) Longipeditermes longipes (47) Nasutitermes columbicus (67) Nasutitermes corniger (67) Nasutitermes ephratae (44, 67, 69, 72) Nasutitermes gagei (73) Nasutitermes nigriceps (67, 158) Nasutitermes rippertii (44)	X-ray analysis of a 15,16-epoxy-3-acetate of 63 (relative configuration) (44) CD measurement (absolute configuration) (44)
64	1,11:4,16:7,16-Cyclized Cembranoid; Longipanoid (1R*,3S*,4S*,7S*,8S*,11R*,12R*,15S*)-Longip-15(17)-en-3-ol	Longipeditermes longipes (45, 47)	

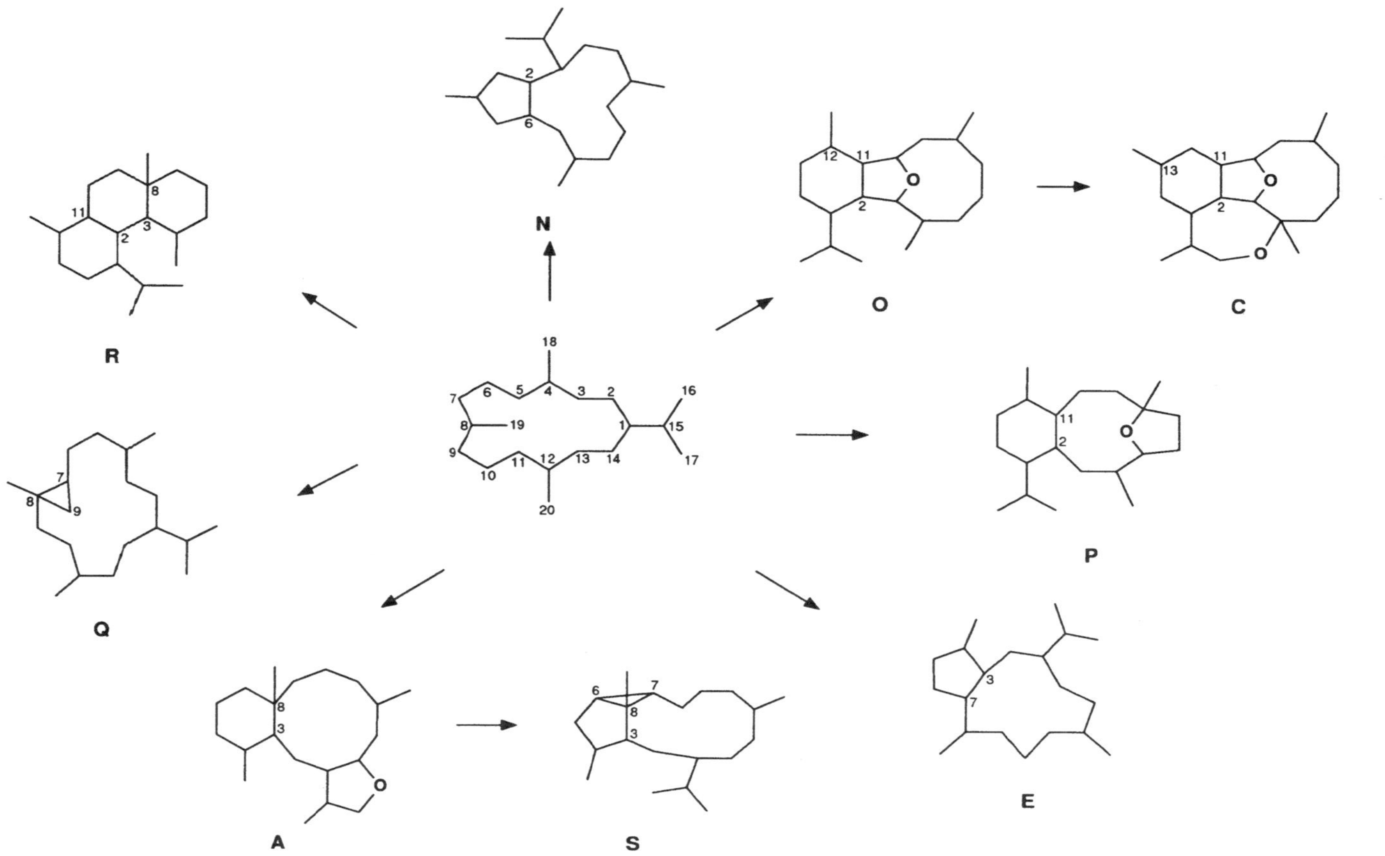

Scheme 5. Proposed biogenic relationship between a cembrane precursor and the 2,6-cyclized cembranoid (**N**), the cladiellins (**O**), the sarcodictyins (**P**), the capnosanoids (**E**), the briarans (**A**), the gersolanoid (**Q**), the 2,11:3,8-cyclized cembranoid (**R**), the asbestinins (**C**) and the erythranoid (**S**)

The gersolane skeleton (**Q**) (*94*) is thought to arise by rearrangement of the cembrane skeleton. Only one compound of this class (**227**) has hitherto been found. There is also only one compound (**228**) having the tricyclic 2,11:3,8-cyclized cembrane skeleton (**R**) (*95*).

The formation of the asbestinins (**C**) (*14*) (**99–109**) is likely to take place *via* methyl migration in a suitable cladiellin precursor or precursors. In addition to the 2,9-epoxide the asbestinins possess a 3,16-epoxy bridge. The erythrane skeleton (**S**), represented by erythrolide A (**226**), is derived from the briaran skeleton by a di-π-methane rearrangement (*96*).

To the best of our knowledge no studies that shed light on the biosynthesis of the cyclized cembranoids mentioned above have been reported. The contention is, however, that cembranoids, which have also been found in many of the marine organisms producing cyclized cembranoids are genuine intermediates (*14, 89, 94, 97*).

It has been suggested that the cyclized cembranoids may serve as defense against predation and against settling by larvae. Consistent with this view, studies have shown that e.g. pteroidine (**209**) and the related **210** are ichthyotoxic (*98*) and that stylatulide (**188**) and the renillafoulins A-C (**139–141**) are active against larval settling (*99–101*).

Reviews covering various aspects of cyclized cembranoids of marine origin have previously been published by FAULKNER (*102–107*), FENICAL (*13*). TURSCH *et al.* (*108*), KREBS (*109*), BAKER and WELLS (*110*) and CARDELLINA (*111*).

A. Gorgonacea

Sea fans and sea whips of the order Gorgonacea have been studied extensively during the past two decades. The results show that these animals elaborate a wide array of cyclized cembranoids comprising briarans, cladiellins, asbestinins, an erythrolane and a tricyclic compound.

1. Astrogorgia, Calicogorgia, Eunicella and Muricella Species

The gorgonian *Eunicella stricta* is the well-known source of eunicillin (**89**), the first compound of the cladiellin type of diterpenoids to be discovered in nature (*1*).

Ophirin (**86**), another cladiellin, has been isolated from a *Muricella* gorgonian of the Red Sea (*93*) and from *Astrogorgia* (*112*) and *Calicogorgia* species (*113*), the latter two collected from Japanese waters. Ophirin (**86**) is an inhibitor of cell division in fertilized starfish eggs as is the structurally closely related astrogorgin (**90**) obtained from the *Astrogorgia* species (*112*). The calicophirins A and B (**91, 74**), which are constituents of the *Calicogorgia* species, are also structural congeners of ophirin (**86**). These two compounds are insect growth inhibitors (*113*).

74 R=H
86 R=OAc
90
91

2. Briareum Species

Corals of the genus *Briareum* are found both in Caribbean and Indo-Pacific waters. They belong to the order Gorgonacea, but their growth and appearance is more similar to the alcyonaceans (*30*). Their chemistry has been examined by several research groups. The results show that they are rich sources of diterpenoids, briarans being the most commonly encountered metabolites. These cooccur with/or are replaced by asbestinins, cladiellins and a 2,11:3,8-cyclized cembranoid in certain species.

Extracts from *Briareum asbestinum*, a species dwelling on reefs in the Caribbean, have been reported to exhibit antitumour activity and toxicity to goldfish (*32, 33*). Consistent with this several of the briarans and asbestinins isolated are noted for their biological activities. Thus, brianthein Z (**193**) and V (**197**) show cytotoxic as well as antiviral effects, while brianthein Y (**194**) is antiviral (*114*). Acetylcholine and histamine antagonism have been reported for the asbestinins **104** and **106** (*32*), and the 4-deoxyasbestinins **99–102** exhibit cytotoxic and antimicrobial activities (*33*).

The briarans so far encountered in *B. asbestinum* (**192–194, 197, 211, 212**) are all chlorinated at C-6 and possess hydroxy or ester substituents at C-2 and C-9. They differ mainly with respect to the substitution of the six-membered ring. Eleven asbestinins (**99–109**) are known to-date, *B. asbestinum* being the sole source (*14, 32, 33*).

Brianthein X, Y and Z (**192–194**) are also metabolites of *B. polyanthes*, a species collected from Bermudian water. Brianthein W (**125**), another constituent of this species, is nonchlorinated and possesses an α,β-unsaturated γ-lactone ring (*35, 115, 116*). This compound (**125**) is structurally closely related to the four briarans (**127, 129–131**) isolated from *B. steckei* of the Barrier Reef (*30*).

In addition to the erythrolides A, B and D (**226, 200, 187**), which have been isolated from *Erythropodium caribaeorum* (*96, 117*), a *Briareum* species of Caribbean origin has yielded nine briarans. These compounds, named briareolide A-I (**171, 170, 163, 162, 160, 159, 132, 155, 153**), are all nonchlorinated and typically oxygenated at C-2, C-9, C-11, C-12 and C-14. Of interest is the fact that briareolides A-E exhibit antiinflammatory effects (*117*). This is also true for brianolide (**218**), a chloro-containing and highly oxygenated briaran isolated from an Okinawan *Briareum* species (*118*).

A *Briareum* species collected from Australian water contains a wide array of cyclized cembranoids. As many as nine briarans (**128, 152, 156, 157, 165–169**), all nonchlorinated, have been isolated. These cooccur with two cladiellins (**83, 84**) and the tricyclic compound **228** (*95*). The latter is most likely a 2,11:3,8-cyclized cembranoid. Support for this view is provided by the fact that an appropriate cembrane precursor, compound **237**, is present in this species. Moreover, the enantiomer of **228** [(+)-**228**] has been obtained as one of the products of the reaction of cembrene (**238**) with formic acid (*119*). When perchloric acid was used as the acidic reagent and cembrene (**238**) or isocembrol (**239**) were the substrates, the related tricyclic compound **240** was formed (*120*). A mechanism for the generation of (−)-**228** is outlined in Scheme 6.

Another collection of *Briareum* corals of Australian origin furnished two halogenated briarans (**180, 207**) (*121*), while an undescribed species classified as belonging to the genus *Solenopodium* [synonym for Indo-Pacific *Briareum* (*121*)] is the source of the solenolides A-F (**196, 195, 219, 220, 175, 134**). Antiinflammatory, antiviral and insecticidal activities are found among these briarans, which are all oxygenated at C-12 and contain a 13,14 double bond or a 13,14-epoxy group (*122*).

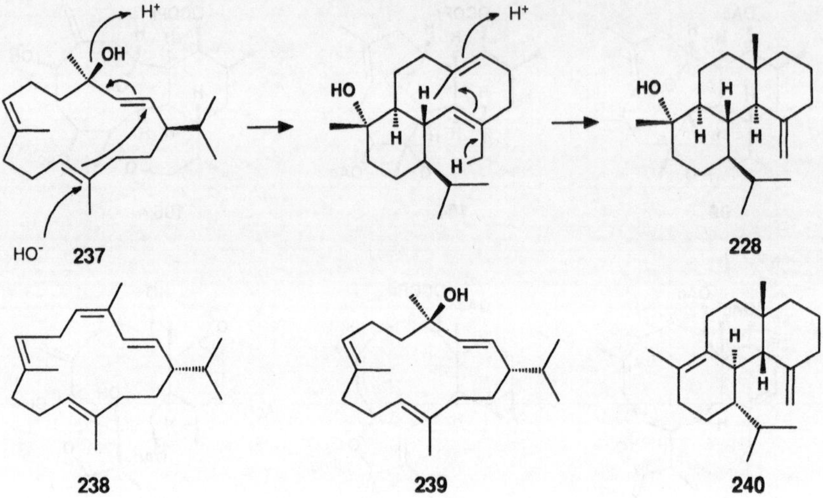

Scheme 6. Conversion of the cembranoid **237** into the 2,11:3,8-cyclized cembranoid **228**. Formulae **238–240**

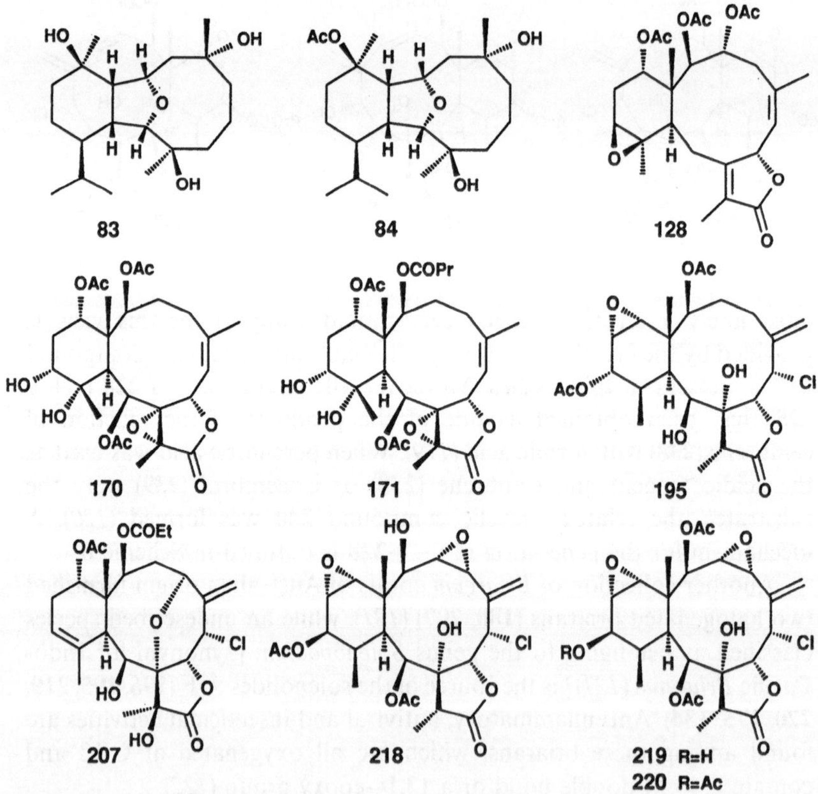

References, pp. 132–141

3. *Erythropodium Species*

In addition to the erythrolides B-I of the briaran class (**200, 186, 187, 202, 203, 201, 149, 204**), the Caribbean octocoral *Erythropodium caribaeorum* contains erythrolide A (**226**), a compound which has a unique carbotricyclic erythrane skeleton. It has been suggested and is supported by experimental results (*96*) that erythrolide A (**226**) is formed from erythrolide B (**200**) by a photochemically induced di-π-methane rearrangement involving the 2,3 and 13,14 double bonds (Scheme 7).

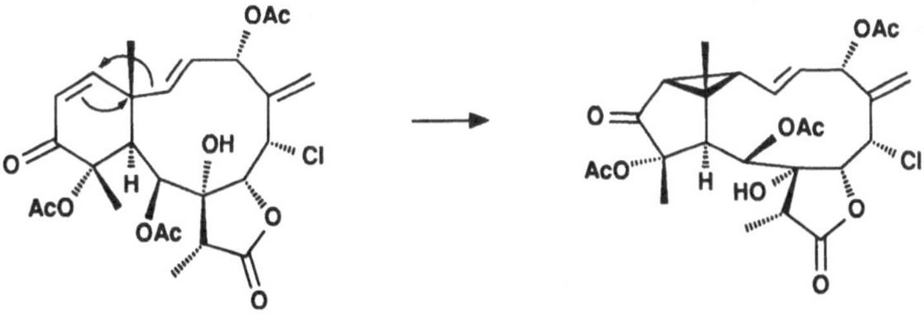

200 → **226**

Scheme 7. Conversion of erythrolide B (**200**) into erythrolide A (**226**) by a di-π-methane rearrangement

The briarans encountered in *E. caribaeorum* are all structurally closely related. All but erythrolide G (**201**) incorporate an α,β-unsaturated oxo group. Erythrolide H (**149**) is unique in being without chlorine and in having a hydroxy substituent at C-16. While the erythrolides, C, D and H (**186, 187, 149**) possess a (2R*,3R*)-epoxide group, a (2S*,8R*)-epoxide group is present in the erythrolides E, F, I and G (**202–204, 201**). It seems plausible therefore that erythrolide C (**186**) may serve as a precursor of erythrolide E (**202**), the conversion involved being an intramolecular attack of the hydroxy group at C-8 on C-2 (Scheme 8) (*117*).

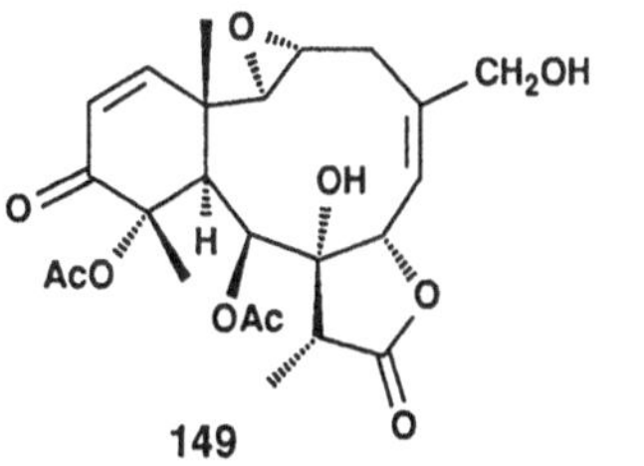

149 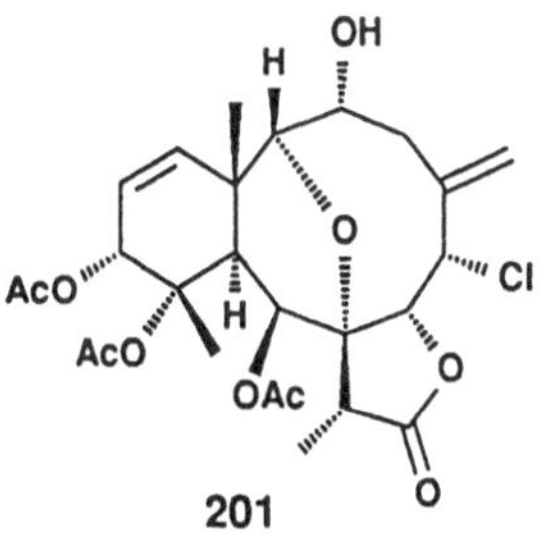   **201**

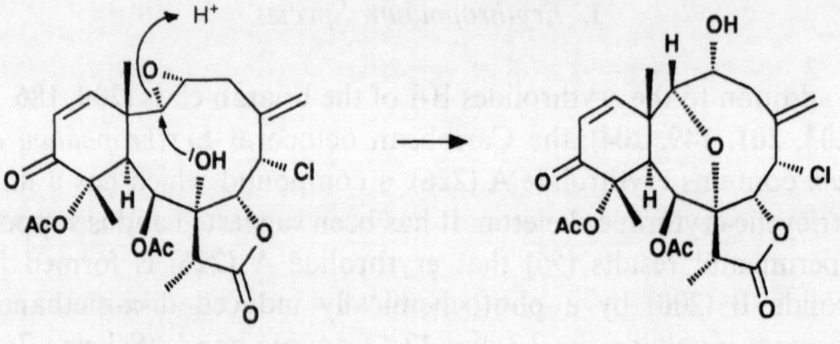

186 202

Scheme 8. Proposed bioconversion of erythrolide C (**186**) into erythrolide E (**202**)

4. *Junceella and Plexaureides Species*

The chemistry of four species of the genus *Junceella*, *J. fragilis*, *J. gemmacea*, *J. juncea* and *J. squamata*, has been examined. All are briaran producers. Junceellin (**205**) and junceellin B (**217**) are metabolites of *J. squamata* from the South China Sea (*123*, *124*). The former compound cooccurs with and is a probable precursor of praeolide (**208**) in *J. fragilis*. Also present in this inhabitant of Chinese water are the junceellolides A-D (**206**, **179**, **199**, **136**). These exhibit antiinflammatory effects (*31*).

Like the aforementioned *Junceella* metabolites, those obtained from *J. juncea* of the Red Sea, i.e. the juncins A-F (**198**, **178**, **214**, **213**, **223**, **216**)

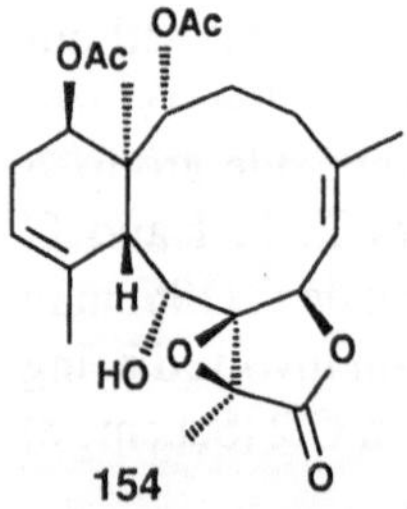

154

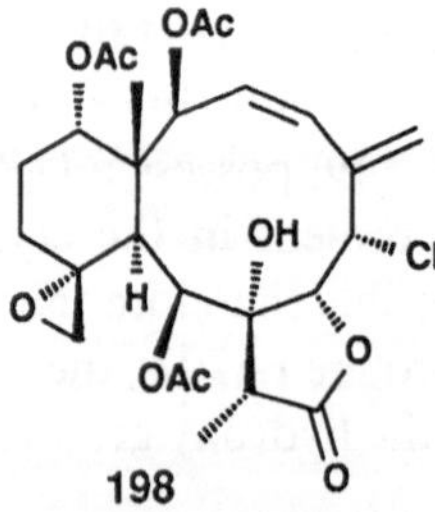

198

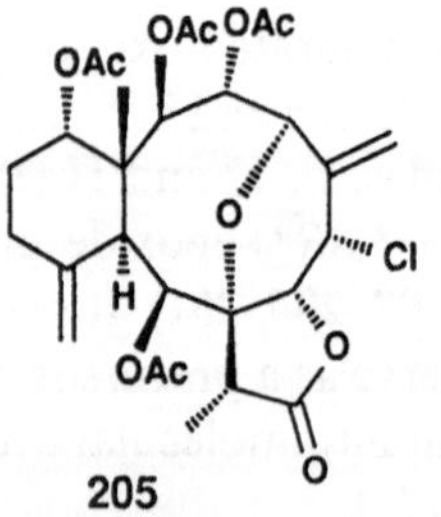

205

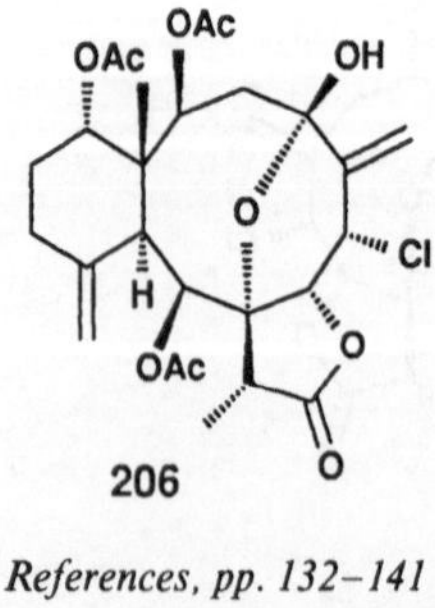

206

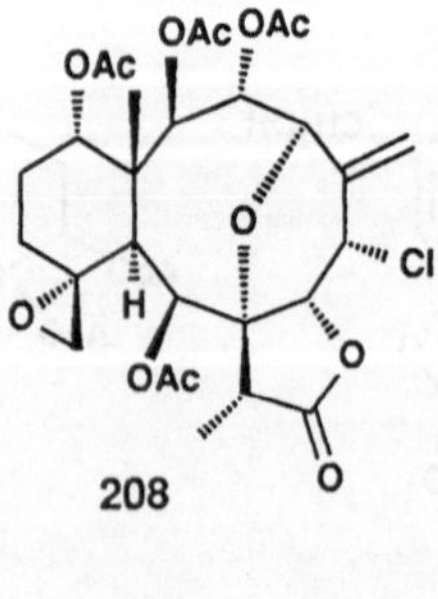

208

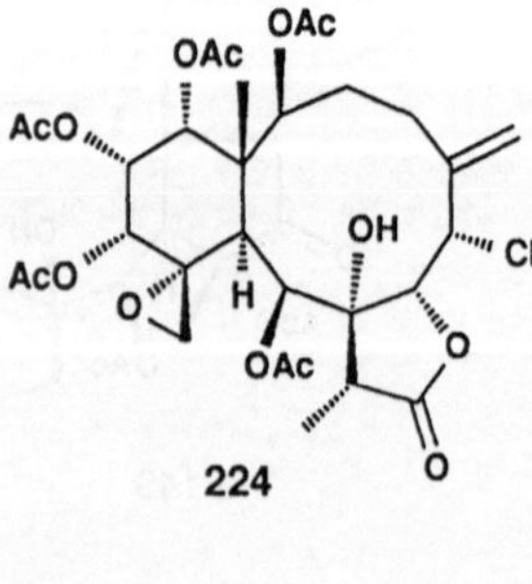

224

possess an 11,20 double bond or an 11,20 epoxide group (*125*). This is also true for all but one (**154**) of the constituents encountered in *J. gemmacea*. These comprise the gemmacolides A-F (**224, 225, 215, 222, 221, 151**) from a Micronesian collection (*126*) and three 8,17-epoxy-containing compounds (**154, 158, 161**) from an Australian collection (*127*). It is noteworthy that the absolute configuration at C-1 and C-10 of the latter three compounds, as determined by Horeau's method, differs from that established for C-1 and C-10 of other briarans.

The 4,8-epoxy bridged briaran praeolide (**208**) mentioned above was first isolated from *Plexaureides praelonga*, a gorgonian dwelling in Chinese water (*128, 129*).

B. Alcyonacea

Soft corals of the order Alcyonacea are known as sources of cladiellins, briarans, capnosanoids, a 2,6-cyclized cembranoid and a gersolane. It is noteworthy that several of these alcyonaceans are well-established cembrane producers as well.

I. Alcyonium Species

Alcyonium coralloides, a Mediterranean soft coral, is the source both of structurally complex cembranolides and of unusual cyclized cembranoids. Of the latter coralloidolide C (**113**) has a capnosanoid skeleton, while coralloidolide F (**65**) is unique in being cyclized across carbons 2 and 6 of the cembrane skeleton (*89, 97*). It has been proposed that coralloidolide C (**113**) may be generated as illustrated in Scheme 9. Coralloidolide E (**241**), a constituent of *A. coralloides*, undergoes hydration of the 7,8 double and enolization involving the oxo group at C-6. Carbon-carbon bond closure between C-3 and C-7 leads to the capnosanoid **242**, which suffers dehydration to form coralloidolide C (**113**) (*97*).

65   **92**

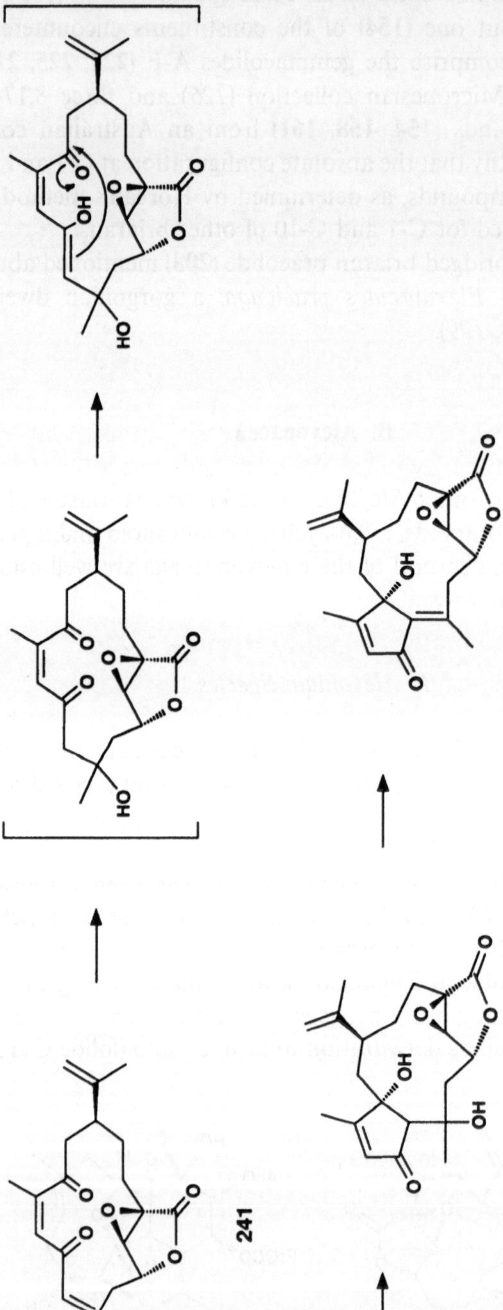

Scheme 9. Proposed biogenesis of coralloidolide C (113) from coralloidolide E (241)

Another *Alcyonium* species, *A. molle* has yielded a highly oxygenated cladiellin-based compound (**92**) (*90*).

2. Cespitularia Species

Two compounds having capnosane type skeletons (**110, 111**) have been isolated from a *Cespitularia* species (*130*). It is interesting to note that species of this genus are known to elaborate cembranoids.

67 R=Ac
68 R=COC₃H₇

75

77

79

110

111

3. Cladiella, Litophyton, Sclerophytum and Sinularia Species

Diterpenoids based on the cladiellin skeleton (**66, 67, 72, 78**) have been discovered in various *Cladiella* species (*131–133*).

The litophynins A-H (**68, 71, 73, 85, 81, 69, 70, 82**), which are all of the cladiellin type, have been discovered as constituents of a *Litophyton* species (*134–137*). Of these, the litophynins A-C and G (**68, 71, 73, 70**) have demonstrated insect growth inhibitory activity.

Sclerophytum capitalis is the source of the sclerophytins A-F (**75, 76, 88, 87, 80, 79**) (*138, 139*). Alcyonin (**77**), another cladiellin, has been isolated from *Sinularia flexibilis* (*140*). Both alcyonin (**77**) and sclerophytin A (**75**) are cytotoxic.

4. Gersemia Species

Gersolide (227), a macrocyclic diterpenoid incorporating a cyclopropane ring, has been found in a collection of *Gersemia rubiformis*, where it cooccurs with structurally reminiscent cembranoids and pseudopteranoids. It has been suggested that gersolide (227) is generated by rearrangement of a cembrane precursor (*94, 141*).

5. Minabea Species

The genus *Minabea* is unusual among the alcyonaceans, since it has been revealed that one of its species elaborates briarans. The compounds identified include the minabeins 1–10 (184, 185, 183, 182, 181, 138, 137, 139, 147, 148), which are all oxygenated at C-11 and C-12, and stylatulide lactone (135) (*15*).

135   184   227

6. Sarcophyton Species

Sarcophytol L (112) has been isolated from *Sarcophyton glaucum* (*142*). This compound, which has a capnosane skeleton may, however, be an artifact formed by autooxidation and rearrangement of sarcophytol A (243), one of the many cembranoids present in this species. KOBAYASHI et al. (*143*) have studied this process and found that the steps outlined in Scheme 10 are involved. Epoxidation of the 3,4 double bond gives the (3S,4S)-epoxide 244, which undergoes isomerization to form the 4-hydroxy-1,14-epoxide 245. A transannular reaction with participation of the 2,3 and 7,8 double bonds completes the generation of sarcophytol L (112).

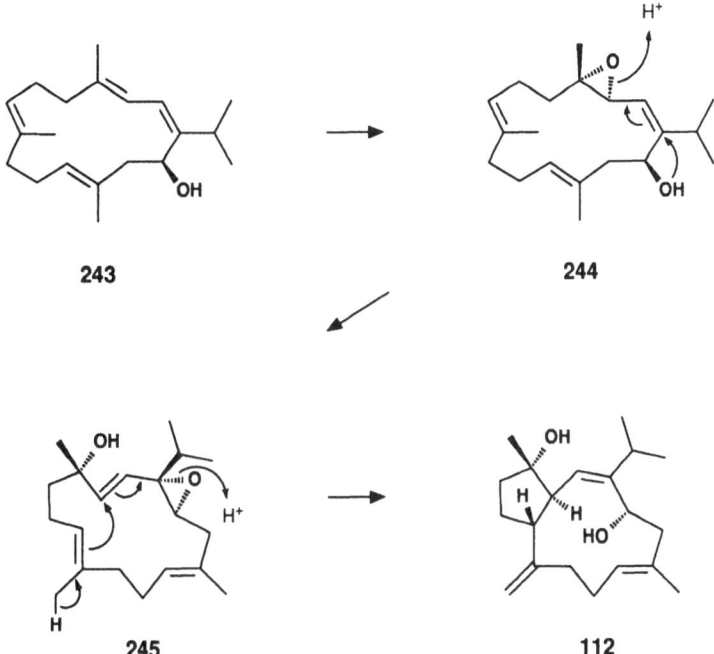

Scheme 10. Conversion of sarcophytol A (**243**) into sarcophytol L (**112**)

C. Stolonifera

Stolonifers are less common than are gorgonians and alcyonaceans in coral reefs. They possess the capability to produce various types of diterpenoids including cyclized cembranoids.

Sarcodictyon roseum, a Mediterranean species, is the source of six unique diterpenoids, the sarcodictyins A-F (**93, 94, 96, 98, 97, 95**). These compounds have the same carbon skeleton as the cladiellins but lack their 2,9-epoxy bridge. They are all hemiacetals and the hydroxy group at C-11 is esterfied by (*E*)- or (*Z*)-*N*(1)-methylurocanic acid (*91, 92*).

The genus *Pachyclavularia* also belongs to the order Stolonifera. In addition to cembranoids (*16*), a species of this genus is known to synthesize acetoxycladiellin (**72**) and briarans (**118, 122, 133**) (*144*). Tubiporein (**164**), isolated from a *Tubipora* species, is a cytotoxic briaran closely related to the minabeins (*145*).

72   93   118

D. Pennatulacea

Several genera of the order Pennatulacea have been examined. The results show that cyclized cembranoids are synthesized by these animals and that briarans are the prevalent compounds.

1. Cavernulina, Pteroides, Ptilosarcus, Renilla, Scytalium and Stylatula Species

Cavernuline (**142**) and its congeners **143** and **144** are present in *Cavernulina grandiflora*. All three compounds are nonchlorinated and possess ester groups at C-2, C-13, and C-14 (*146*). Of the briarans discovered in *Pteroides laboutei*, pteroidine (**209**) and the related **210** are hemiacetals and contain chlorine at C-6, the remaining metabolite labouteine (**150**) being devoid of chlorine (*98*).

Ptilosarcus gurneyi, another sea pen, is the source of ptilosarcone (**191**), ptilosarcenone (**176**) and five other structurally closely related briarans (**172–174, 184, 190**). Of these, **184** is identical with minabein-1 (*105*). Insecticidal effects are found among these compounds (*34, 147*). Worthy of mention is the isolation of ptilosarcenone (**176**) and its butanoate analogue (**177**) from *Tochuina tetraquetra*, a nudibranch whose diet includes *P. gurneyi* (*148*).

The renillafoulins A-C (**139–141**), which inhibit the settlement of barnacle larvae, are produced by the sea pansy *Renilla reniformis*. All three compounds are structural congeners characterized by the presence of an α,β-unsaturated oxo group system in the six-membered ring of the briaran skeleton (*101*). It should be pointed out that renillafoulin A (**139**) is identical with minabein-8 (*105*).

The Australian sea pen *Scytalium tentaculatum* is the source of three briarans (**120, 124, 126**), of which the former two incorporate a furan in place of the γ-lactone moiety (*29*).

The briarans isolated from a *Stylatula* species from the Gulf of California include the toxin stylatulide (**188**) and four related compounds (**189, 135, 145, 146**). All five possess an 11,12 double bond or an 11,12-epoxy group (*99, 100*).

120	139	142

176 R=Ac 177 R=COC₃H₇	188	209

2. Veretillum Species

A series of briarans, the verecynarmins A-G (**119, 114, 116, 123, 117, 115, 121**) has been detected in *Veretillum cynomorium*, a pennatulacean coral of the Mediterranean Sea, and also in its predator the mollusc *Armina maculata* (*149–151*). These compounds incorporate a furan moiety and are nonchlorinated at C-6. Verecynarmin D (**123**) is, however, unique in having a chlorine substituent at C-13 of the six-membered ring.

119	123

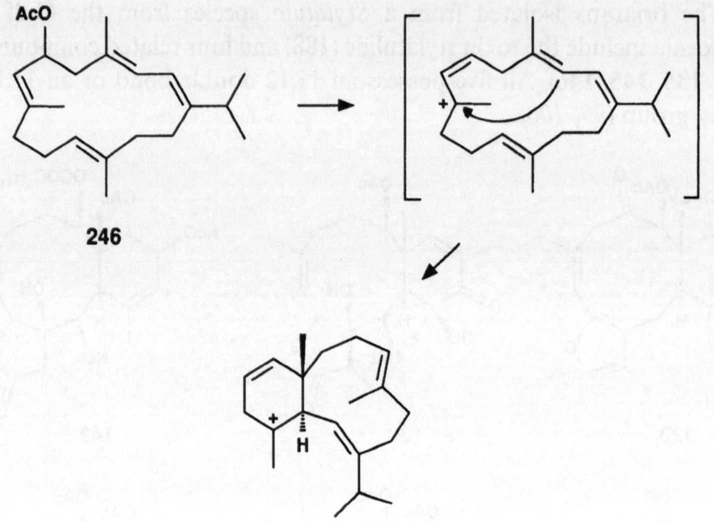

Scheme 11. Proposed biogenesis of preverecynarmin (**246**) into a briarane precursor of the verecynarmins A-G

Noteworthy is the discovery of the cembranoid preverecynarmin (**246**) in *V. cynomorium*. A hypothetical route involving cyclization of this cembranoid to form a briarane precursor of the verecynarmins A-G has been formulated and is shown in Scheme 11 (*151*).

Table 3. Cyclized Cembranoids from Marine Organisms

| Structure no. | Compound | | Source | X-Ray Study | Biological Activity |
	Name				

2,6-Cyclized Cembranoid

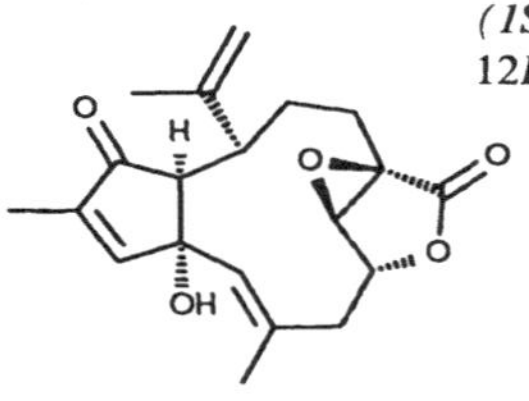

(1S*,2S*,6R*,7E,10R*,11R*, 12R*)-Coralloidolide F

Alcyonium coralloides (89)

65

2,11-Cyclized Cembranoids; Cladiellins

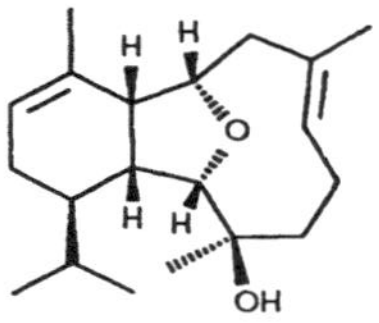

(1R*,2R*,3R*,6E,9R*,10R*, 14R*)-Cladiella-6,11-dien-3-ol

Cladiella species (132)

66

Table 3 (continued)

Structure no.	Compound Name	Source	X-Ray Study	Biological Activity
67	(1R*,2R*,3R*,6E,9R*,10R*, 14R*)-Cladiellin	Cladiella species	(131)	Antiinflammatory effect (110)
68	(1R*,2R*,3R*,6E,9R*10R*, 14R*)-Litophynin A	Litophyton species	(134)	Inhibitory activity against the silkworm Bombyx mori L: ED$_{50}$ 12 ppm (134)
69	(1R*,2R*,3R*,6S*,9R*,10R*, 14R*)-Litophynin F	Litophyton species	(137)	

Inhibitory activity against the silkworm *Bombyx mori* L: ED$_{50}$ 42 ppm *(137)*

(137)

Litophyton species

(1R*,2R*,3R*,9R,10R*, 14R*)-Litophynin G

70

Inhibitory activity against the silkworm *Bombyx mori* L: ED$_{50}$ 2.7 ppm *(134)*

(134)

Litophyton species

(1R*,2R*,3R*,6E,8R*,9S*, 10R*,14R*)-Litophynin B

71

X-ray analysis of **72** (relative configuration) *(131)*

(131)
(144)

Cladiella species
Pachyclavularia species

(1R*,2R*,3R*,6E,9R*,10S*, 11R*,14R*)-Acetoxycladiellin

72

Table 3 (*continued*)

Structure no.	Compound Name	Source	X-Ray Study	Biological Activity
73	(1R,2R,3R,6E,9R,10R,12S,14R)-Litophynin C	*Litophyton* species	(135)	Inhibitory activity against the silkworm *Bombyx mori* L: ED$_{50}$ 25 ppm (135)
74	(1R*,2R*,3R*,6E,9R*,10R*,13S*,14R*)-Calicophirin B	*Calicogorgia* species	(113)	Inhibitory activity against the silkworm *Bombyx mori* L: ED$_{50}$ 52 ppm (113)
75	(2R*,3R*,6S*,7R*,9R*)-Sclerophytin A	*Sclerophytum capitalis*	(138, 139)	Cytotoxic activity against the L1210 cell line at 0.001 µg/ml (138)

(2R*,3R*,6S*,7R*,9R*)-
Sclerophytin B

Sclerophytum capitalis

(138, 139)

76

(1R*,2R*,3R*,4S*,6S*,9R*,
10R*,14R*)-Alcyonin

Simularia flexibilis

(140)

Cytotoxic activity
against Vero cells:
IC$_{50}$ 55 µg/ml (140)

77

(1R*,2R*,3R*,6S*,7S*,9R*,
10R*,14R*)-Cladiell-11-ene-
3,6,7-triol

Cladiella species

(133)

X-ray analysis
of **78** (relative
configuration)
(133)

78

(1R*,2R*,3R*,6S*,7S*,9R*,
10R*,14R*)-Sclerophytin F

Sclerophytum capitalis

(139)

79

Table 3 *(continued)*

Structure no.	Compound Name	Source	X-Ray Study	Biological Activity
80	$(1R^*,2R^*,3R^*,6S^*,7S^*,9R^*, 10R^*,14R^*)$-Sclerophytin E	*Sclerophytum capitalis*	*(139)*	
81	$(1R^*,2R^*,3R^*,6S^*,7R^*,9R^*, 10R^*,14R^*)$-Litophynin E	*Litophyton* species	*(136)*	
82	$(1R^*,2R^*,3R^*,6S^*,9R^*,10R^*, 12S^*,14R^*)$-Litophynin H	*Litophyton* species	*(137)*	

83 (1R*,2R*,3R*,7R*,9R*,10S*,11R*,14R*)-Cladiellane-3,7,11-triol — Briareum species — (95)

84 (1R*,2R*,3R*,7R*,9R*,10S*,11R*,14R*)-11-Acetoxycladiellane-3,7-diol — Briareum species — (95)

85 (1R,2R,3R,6E,9R,10R,12R,13S,14R)-Litophynin D — Litophyton species — (136) — Exhibits brine shrimp lethality LD_{50} 0.9 ppm (136)

86 (1S*,2R*,3R*,6E,9R*,10R*,13R*,14R*)-Ophirin — Astrogorgia species, Calicogorgia species, Muricella species — (112) (113) (93) — Inhibitor of cell division in fertilized starfish eggs at 10 µg/ml (112)

Table 3 (continued)

Structure no.	Compound Name	Source	X-Ray Study	Biological Activity
87	(1R*,2R*,3R*,6S*,7R*, 8S*,9S*,10R*,14R*)- Sclerophytin D	Sclerophytum capitalis	(139)	
88	(1R,2R,3R,6S,7R,8S,9S, 10R,14R)-Sclerophytin C	Sclerophytum capitalis	(139) X-ray analysis of 88 (absolute configuration) (139) MM2 calculations (139)	
89	(1R*,2R*,3R*,6S*,9R*,10S*, 11S*,12S*14R*)-Eunicellin	Eunicella stricta	(1) X-ray analysis of the dibromide of 89 (relative configuration) (1, 152)	

90

(1S*,2R*,3R*,6S*,9R*,10R*, 13R*,14R*)-Astrogorgin

Astrogorgia species

(112)

Inhibitor of cell division in fertilized starfish eggs at 10 μg/ml (112)

91

(1R*,2R*,3R*,6E,9R*,10S*, 11S*,12R*,13S*,14R*)-Calicophirin A

Calicogorgia species

(113)

Inhibitory activity against the silkworm *Bombyx mori* L: L: ED_{50} 20 ppm (113)

92

(1S,2S,3S,4S,6E,9S,10R,11R, 12S,13R,14S)-12-Acetoxy-3,13-dibutanoyloxycladiell-6-ene-4,11-diol

Alcyonium molle

(90)

Table 3 (*continued*)

Structure no.	Compound Name	Source	X-Ray Study	Biological Activity
	2,11-Cyclized Cembranoids: Sarcodictyins			
 93	(4*R*,4a*R*,5*E*,7*R*,10*S*,11*S*, 12a*R*)-Sarcodictyin A	*Sarcodictyon roseum*	(*91*)	
 94	(4*R*,4a*R*,5*E*,7*R*,10*S*,11*S*, 12a*R*)-Sarcodictyin B	*Sarcodictyon roseum*	(*91*)	

(1R*,4R*,4aR*,5E,7R*,10S*,11S*,12aR*)-Sarcodictyin F

Sarcodictyon roseum (92)

95

(3R*,4S*,4aS*,5E,7S*,10R*,11R*,12aS*)-Sarcodictyin C

Sarcodictyon roseum (92)

96

(3R*,4S*,4aS*,5E,7S*,10R*,11R*,12aS*)-Sarcodictyin E

Sarcodictyon roseum (92)

97

Table 3 (continued)

Structure no.	Compound Name	Source	X-Ray Study	Biological Activity
98	(3R*,4S*,4aS*,5E,7S*,10R*, 11R*,12aS*)-Sarcodictyin D	*Sarcodictyon roseum*	(92)	
2,11-Cyclized Cembranoids; Asbestinins 99	(1S*,2S*,3S*,6E,9S*,10S*, 11R*,12R*,14S*,15R*)-11-Acetoxy-4-deoxyasbestinin B	*Briareum asbestinum*	(33)	Cytotoxic effect against CHO-K1 cells: ED$_{50}$ 2.50 µg/ml. Strong antimicrobial activity against *Klebsiella pneumoniae* (33)

100

$(1S*,2S*,3S*,6Z,9S*10S*,11R*,12R*,14S*,15R*)$-11-Acetoxy-4-deoxyasbestinin D

Briareum asbestinum

(33)

Cytotoxic effect against CHO-K1 cells: ED_{50} 4.82 µg/ml. Strong antimicrobial activity against *Klebsiella pneumoniae* (33)

101

$(1S*,2S*,3S*,6E,9S*,10S*,11R*,12R*,14S*,15R*)$-4-Deoxyasbestinin A

Briareum asbestinum

(33)

Cytotoxic effect against CHO-K1 cells: ED_{50} 3.55 µg/ml. Strong antimicrobial activity against *Klebsiella pneumoniae* (33)

102

$(1S*,2S*,3S*,6Z,9S*,10S*,11R*,12R*,14S*,15R*)$-4-Deoxyasbestinin C

Briareum asbestinum

(33)

Cytotoxic effect against CHO-K1 cells: ED_{50} 3.55 µg/ml. Strong antimicrobial activity against *Klebsiella pneumoniae* (33)

Table 3 (continued)

Structure no.	Compound Name	Source	X-Ray Study	Biological Activity
103	(1S*,2S*,3S*,4R*,6E,9S*, 10S*,11R*,12R*,14S*,15R*)- Asbestinin-3	*Briareum asbestinum* (14)		
104	(1S*,2S*,3S*,4R*,6E,9S*, 10S*,11R*,12R*,14S*,15R*)- -Asbestinin-1	*Briareum asbestinum* (14, 32)	X-ray analysis of the diol derived from **104** (relative configuration) (14, 32)	Antagonizes the effects of acetylcholine on guinea pig ileum preparation at a 13% level at a concentration of 16 µg/ml (32)
105	(1S*,2S*,3S*,4R*,6Z,9S*, 10S*,11R*,12R*,14S*,15R*)- Asbestinin-2	*Briareum asbestinum* (14, 32)		

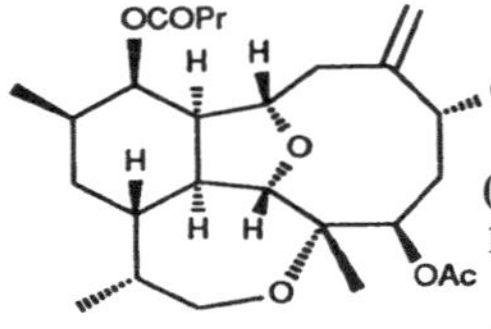

106

Asbestinin-5

(1S*,2S*,3S*,4R*,6R*,9S*,
10S*,11R*,12R*,14S*,15R*)-

Briareum asbestinum

(14, 32)

Exhibits histamine antagonism at a level of 40% at a concentration of 16 μg/ml.
Antagonizes the effects of acetylcholine on guinea pig ileum preparations at a 38% level at a concentration of 16 μg/ml (32)

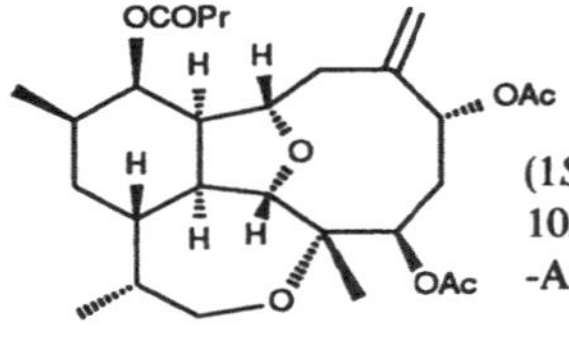

107

-Asbestinin-5 acetate

(1S*,2S*,3S*,4R*,6R*,9S*,
10S*,11R*,12R*,14S*,15R*)

Briareum asbestinum

(32)

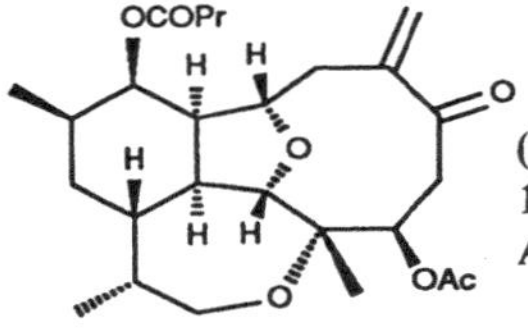

108

Asbestinin-4

(1S*,2S*,3S*,4R*,9S*,10S*,
11R*,12R*,14S*,15R*)-

Briareum asbestinum

(14, 32)

Table 3 (continued)

Structure no.	Compound Name	Source	X-Ray Study	Biological Activity
109	(1S*,2S*,3S*,4R*,6S*,7R*, 9S*,10S*,11R*,12R*,14S*,15 R*)-Asbestinin epoxide	Briareum asbestinium	(32)	
3,7-Cyclized Cembranoids; Capnosanoids				
110	(1E,8E,11E)-Capnosa-1,8,11-trien-4-ol	Cespitularia species	(130)	
111	(1E,3R*,4R*,7S*,8S*,11E)-Capnosa-1,11-diene-4,8,-diol	Cespitularia species	(130)	X-ray analysis of **111** (relative configuration) (130)

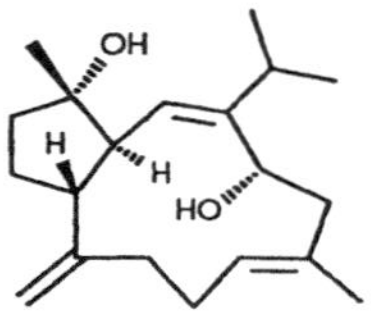

Sarcophyton glaucum (142)

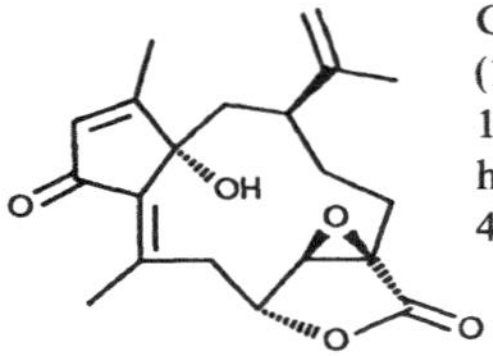

Alcyonium coralloides (97)

Armina maculata (150)
Veretillum cynomorium (150)

Sarcophytol L
(1Z,3R,4S,7R,11E,14S)-
Capnosa-1,8(19),11-triene-
4,14-diol

Coralloidolide C
(1S*,3S*,7E,10R*,11R*,
12R*)-11,12-Epoxy-3-
hydroxy-6-oxocapnosa-
4,7,15-trien-20,10-olide

3,8-Cyclized Cembranoids: Briarans

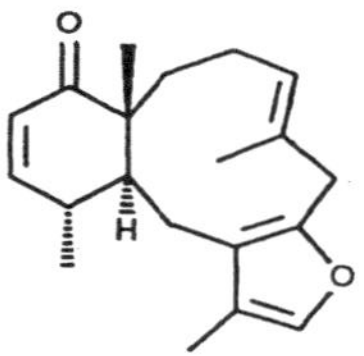

(1R*,4E,10S*,11R*)-
Verecynarmin B

112

113

114

Table 3 (continued)

Structure no.	Compound Name	Source	X-Ray Study	Biological Activity
115	(1R*,5Z,10S*,11R*) Verecynarmin F	Armina maculata Veretillum cynomorium	(151) (151)	
116	(1R*,4E,10R*,11S*)- Verecynarmin C	Armina maculata Veretillum cynomorium	(150) (150)	
117	(1R*,5Z,10R*,11S*)- Verecynarmin E	Armina maculata Veretillum cynomorium	(151) (151)	

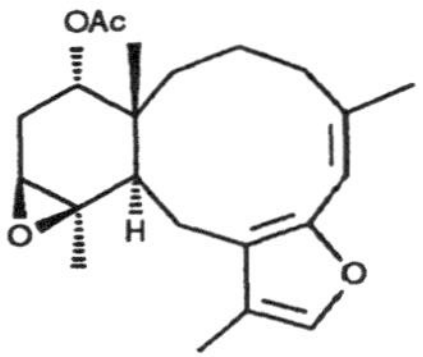

(1R*,5Z,10R*,11R*,12S*,14S*)-11,12-Epoxybriara-5,7,17-trien-14-yl acetate

Pachyclavularia species

(144)

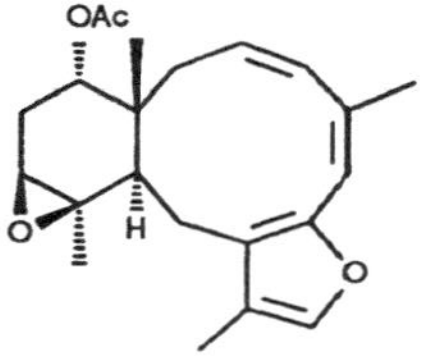

118

(1R,3Z,5Z,10R,11R,12S,14S)-Verecynarmin A

Armina maculata
Veretillum cynomorium

(149)
(149)

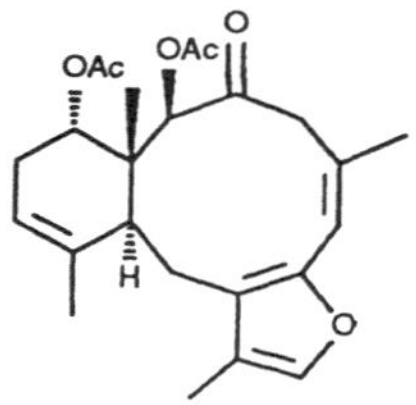

119

(1R*,2R*,5Z,10S*,14S*)-2,14-Diacetoxybriara-5,7,11,17-tetraen-3-one

Scytalium tentaculatum

(29)

Increases the blood pressure and heart beat in cats at a concentration of 150 mg/kg, i.p. (29)

120

Table 3 (continued)

Structure no.	Compound Name	Source	X-Ray Study	Biological Activity
121	(1R*,4S*,5Z,10R*,11R*, 12S*,14S*)-Verecynarmin G	*Armina maculata* *Veretillum cynomorium*	(151) (151)	
122	(1R*,4S*,5Z,10R*,11R*, 12S*,14S*)-11,12-Epoxybriara-5,7,17-triene-4,14-diyl diacetate	*Pachyclavularia* species	(144)	Ichthyotoxic effect on the fish *Gambusia affinis* at a concentration of 2-5 ppm (144)

(1R*,4E,10R*,11R*)-
Verecynarmin D

Armina maculata
Veretillum cynomorium

(150)
(150)

123

(1R*,2R*,5Z,10R*,11S*,
12R*,14S*)-14-Acetoxy-
11,12-epoxy-2-(3-
methylbutanoyloxy)-briara-
5,7,17-trien-3-one

Scytalium tentaculatum

(29)

124

(1S*,2S*,5Z,7S*,10S*,14S*)-
Brianthein W

Briareum polyanthes

(116)

X-ray analysis
of **125**
(relative
configuration)
(116)

125

Table 3 (continued)

Structure no.	Compound Name	Source	X-Ray Study	Biological Activity
126	(1S*,2S*,6E,10S*,14S*)-14-Acetoxy-5-hydroxy-2-(3-methylbutanoyloxy)-briara-6,8(17),11-trien-18-one	Scytalium tentaculatum	(29)	
127	(1S*,2S*,5Z,7S*,10R*,11R*,12S*,14S*)-14-Acetoxy-2-butanoyloxy-11,12-epoxy-briara-5,8(17)-dien-18-one	Briareum steckei	(30)	

X-ray analysis
of **130**
(relative
configuration)
(*30*)

(*30*)

(*30*)

(*95*)

Briareum species

Briareum steckei

Briareum steckei

(1*R**,2*R**,3*R**,5*Z*,7*S**,
10*R**,11*R**,12*S**,14*S**)-
2,3,14-Triacetoxy-11,12-
epoxybriara-5,8(17)-dien-18-
one

(1*R**,2*R**,3*R**,5*Z*,7*S**,10*R**,
11*R**,12*S**,14*S**)-2-
Butanoyloxy-3,14-diacetoxy-
11,12-epoxybriara-5,8(17)-
dien-18-one

(1*S**,2*S**,4*R**,5*Z*,7*S**,10*R**,
11*R**,12*S**,14*S**)-4,14-
Diacetoxy-11,12-epoxy-2-
propanoyloxybriara-5,8(17)-
dien-18-one

128

129

130

Table 3 (*continued*)

Structure no.	Compound Name	Source	X-Ray Study	Biological Activity
131	(1S*,2S*,4R*,5Z,7S*,10R*, 11R*,12S*,14S*)-2-Butanoyloxy-4,14-diacetoxy-11,12-epoxybriara-5,8(17)-dien-18-one	*Briareum steckei*	X-ray analysis of **131** (relative configuration) (*30*)	
		(*30*)		
132	(1S*,2S*,5Z,7S*,9S*,10S*, 11R*,12R*,14S*)-Briareolide G	*Briareum species*		
		(*117*)		

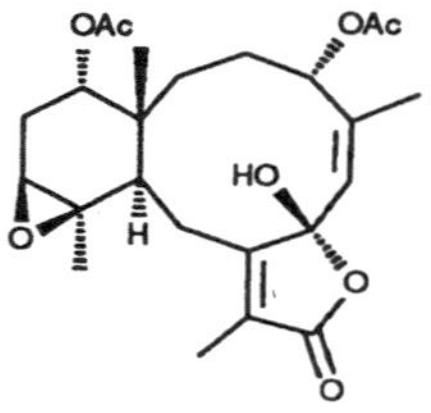

Pachyclavularia species (144)

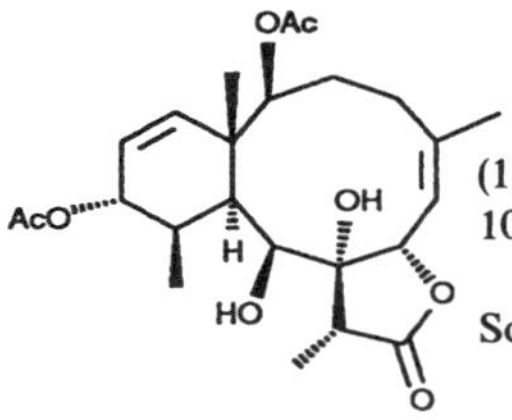

(1R*,4S*,5Z,7S*,10R*,11R*,12S*,14S*)-4,14-Diacetoxy-11,12-epoxy-7-hydroxy-briara-5,8(17)-dien-18-one

133

Solenopodium species (122)

Active against
Blowfly larve:
ED$_{100}$ 30–35 ppm.
Antiinflammatory
properties (122)

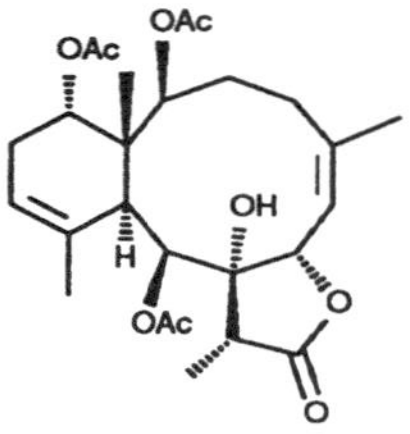

(1S*,2S*,5Z,7S*,8R*,9S*,10S*,11R*,12R*,17R*)-Solenolide F

134

Minabea species (15)
Stylatula species (100)

(1S*,2S*,5Z,7S*,8R*,9S*,10S*,14S*,17R*)-Stylatulide lactone

135

Table 3 (continued)

Structure no.	Compound Name	Source	X-Ray Study	Biological Activity
136	(1R*,2S*,4R*,5Z,7S*,8R*, 9S*,10S*,14S*,17R*)- Juncellolide D	Junceella fragilis	(31)	Reduction (82.3%) of edema in the mouse ear assay (31)
137	(1S*,2S*,5Z,7S*,8R*,9S*, 10S*,11S*,12R*,17R*)- Minabein-7	Minabea species	(15)	

(1S*,2S*,5Z,7S*,8R*,9S*,10S*,11S*,12R*,17R*)-Minabein-6

138

Minabea species

(15)

(1S*,2S*,5Z,7S*,8R*,9S*,10S*,11S*,17R*)-Minabein-8 (Renillafoulin A)

139

Renilla reniformis
Minabea species

(101)
(15)

Inhibits the settlement of larvae of the barnacle *Balanus amphitrite*: EC₅₀ 0.02–0.2 μg/ml (101)

Inhibits the settlement of larvae of the barnacle *Balanus amphitrite*: EC_{50} 0.02–0.2 μg/ml (101)

(1S*,2S*,5Z,7S*,8R*,9S*,10S*,11S*,17R*)-Renillafoulin B

140

Renilla reniformis

(101)

Inhibits the settlement of larvae of the barnacle *Balanus amphitrite amphitrite*: EC_{50} 0.02–0.2 μg/ml (101)

Table 3 (continued)

Structure no.	Compound Name	Source	X-Ray Study	Biological Activity
141	(1S*,2S*,5Z,7S*,8R*,9S* 10S*,11S*,17R*)-Renillafoulin C	Renilla reniformis (101)	X-ray analysis of 141 (relative configuration) (101)	Inhibits the settlement of the larvae of the barnacle Balanus amphitrite amphitrite: EC_{50} 0.02–0.2 µg/ml (101)
142	(1S*,2S*,5Z,7S*,8R*,9S*, 10S*,13R*,14R*,17R*)-Cavernuline	Cavernulina grandiflora (146)		

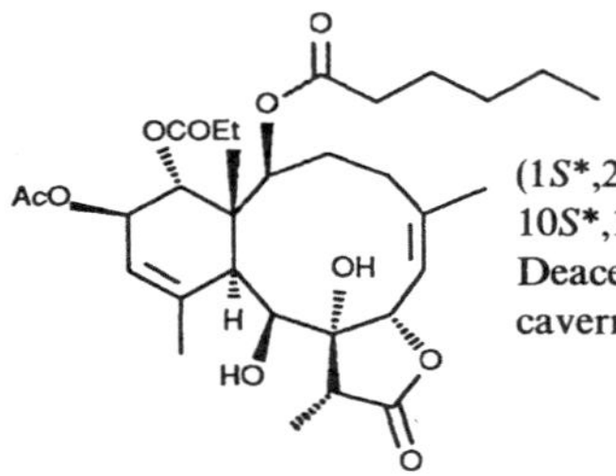

Cavernulina grandiflora   *(146)*

(1S*,2S*,5Z,7S*,8R*,9S*, 10S*,13R*,14R*,17R*)-0-Deacetyl-propionyl-cavernuline

143

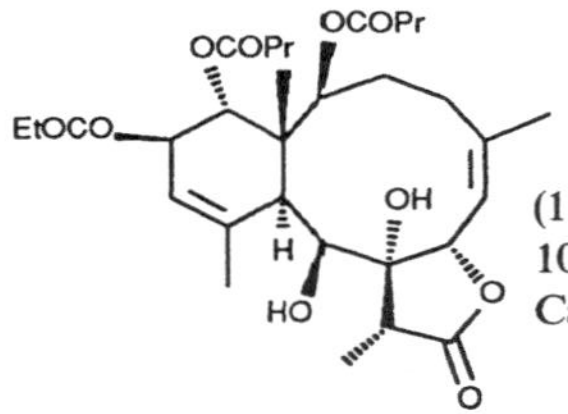

Cavernulina grandiflora   *(146)*

(1S*,2S*,5Z,7S*,8R*,9S*, 10S*,13R*,14R*,17R*)-Cavernulinine

144

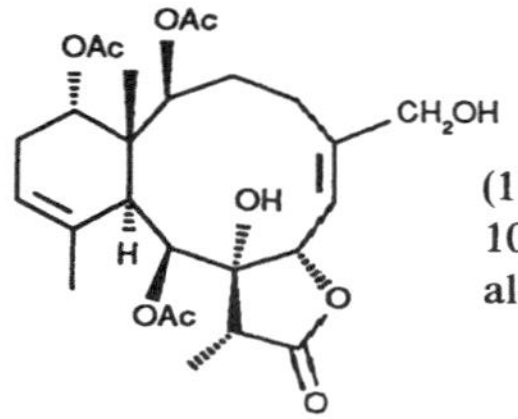

Stylatula species   *(100)*

(1S*,2S*,5E,7S*,8R*,9S*, 10S*,14S*,17R*)-Stylatulide alcohol

145

Table 3 (continued)

Structure no.	Compound Name	Source	X-Ray Study	Biological Activity
146	(1S*,2S*,5E,7S*,8R*,9S*, 10S*,14S*,17R*)-Stylatulide methyl ester	Stylatula species	(100)	
147	(1S*,2S*,5Z,7S*,8R*,9S*, (10S*,11S*,14S*,17R*)- Minabein-9	Minabea species	(15)	

(15)

Minabea species

(1R*,2S*,3Z,5E,7S*,8R*, 9S*,10S*,11S*,12R*,14S*, 17R*)-Minabein-10

148

(117)

Erythropodium caribaeorum

(1S*,2R*,3R*,5E,7S*,8R*, 9S*,10S*,11S*,17R*)- Erythrolide H

149

(98)

Pteroides laboutei

(1R*,2S*,3Z,5E,7S*,8R*, 9S*,10S*,11S*,12R*,14S*, 17R*)-Labouteine

150

Table 3 (continued)

Structure no.	Compound Name	Source	X-Ray Study	Biological Activity

(1S*,2S*,3Z,5E,7S*,8R*,
9S*,10S*,11R*,12R*,13R*,
14R*,17R*)-Gemmacolide F

151

Junceella gemmacea

(126)

(1S*,2S*,5Z,7S*,8R*,9S*,
10S*,11R*,12R*,17R*)-
2,12-Diacetoxy-8,17-epoxy-9-
hydroxybriara-5,13-dien-18-one

152

Briareum species

(95)

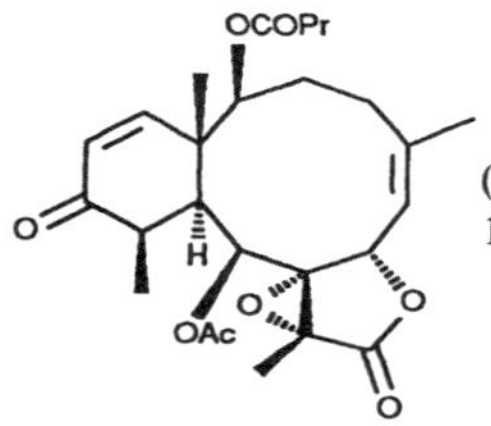

(117)

Briareum species

(1S*,2S*,5Z,7S*,8R*,9S*,
10S*,11R*,17R*)-Briareolide I

153

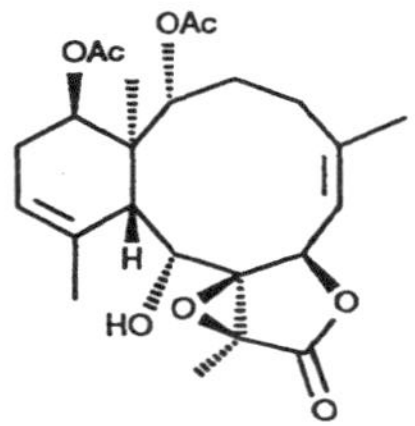

(127)

Junceella gemmacea

(1R,2R,5Z,7R,8S,9R,10R,
14R,17S)-2,14-Diacetoxy-
8,17-epoxy-9-hydroxybriara-
5,11-dien-18-one

154

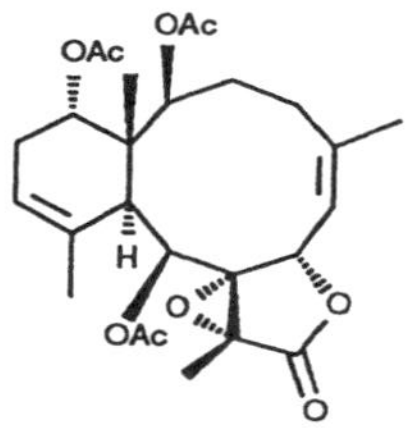

(117)

Briareum species

(1S*,2S*,5Z,7S*,8R*,9S*,
10S*,14S*,17R*)-
Briareolide H

155

Table 3 (continued)

Structure no.	Compound Name	Source	X-Ray Study	Biological Activity
156	(1R*,2R*,3S*,5Z,7S*,8R*, 9S*,10S*,14S*,17R*)-2,3,14-Triacetoxy-8,17-epoxy-9-hydroxybriara-5,11-dien-18-one	Briareum species	(95)	
157	(1R*,2R*,3S*,5Z,7S*,8R*, 9S*,10S*,14S*,17R*)-2,14-Diacetoxy-3-butanoyloxy-8,17-epoxy-9-hydroxybriara-5,11-dien-18-one	Briareum species	(95)	

Inhibition (46%) of inflammation in the mouse ear assay at a dose of 50 μg/ear (*117*)

(*127*)

Junceella gemmacea

(*117*)

Briareum species

(*117*)

Briareum species

(1*R*,2*R*,5*Z*,7*R*,8*S*,9*R*,10*R*, 12*R*,14*R*,17*S*)-2,14-Diacetoxy-8,17-epoxy-9,12-dihydroxybriara-5,11(20)-dien-18-one

(1*S*,2*S*,5*Z*,7*S*,8*R*,9*S*, 10*S*,11*R*, 12*R*,14*S*,17*R*)-Briareolide F

(1*S*,2*S*,5*Z*,7*S*,8*R*,9*S*, 10*S*,11*R*, 12*R*,14*S*, 17*R*)-Briareolide E

158

159

160

Table 3 (continued)

Structure no.	Compound Name	Source	X-Ray Study	Biological Activity
161	(1R,2R,5Z,7R,8S,9R,10R, 11R,14R,17S)-2,14-Diacetoxy-8,17:11,20-diepoxy-9-hydroxybriar-5-en-18-one	Junceella gemmacea	(127)	
162	(1S*,2S*,5Z,7S*,8R*,9S*, 10S*,11S*,12R*,14S*,17R*)-Briareolide D	Briareum species	(117)	Inhibition (85%) of inflammation in the mouse ear assay at a dose of 50 µg/ear (117)

Inhibition (75%) of inflammation in the mouse ear, assay at a dose of 50 μg/ear (117)

(117)

Briareum species

(1S*,2S*,5Z,7S*,8R*,9S*,10S*,11S*,12R*,14S*,17R*)-Briareolide C

163

Cytotoxic against B16 mouse melanoma cells: IC$_{50}$ 2.0 μg/ml (145)

(145)

Tubipora species

(1S*,2R*,3R*,5Z,7S*,8R*,9R*,10S*,11S*,12R*,17R*)-Tubiporein

164

(95)

Briareum species

(1R*,2R*,3R*,5Z,7S*,8R*,9S*,10S*,11R*,12S*,14S*,17R*)-2,3,14-Triacetoxy-8,17:11,12-diepoxy-9-hydroxybriar-5-en-18-one

165

Table 3 *(continued)*

Structure no.	Compound Name	Source	X-Ray Study	Biological Activity
166	(1R*,2R*,3R*,5Z,7S*,8R*, 9S*,10S*,11R*,12S*,14S*, 17R*)-3,14-Diacetoxy-2-butanoyloxy-8,17:11,12-diepoxy-9-hydroxybriar-5-en-18-one	*Briareum* species *Solenopodium stechei*	(95) (159)	
167	(1S*,2S*,4R*,5Z,7S*,8R*, 9S*,10S*,11S*,12R*,17R*)-2,4,9-Triacetoxy-8,17-epoxy-11,12-dihydroxybriara-5,13-dien-18-one	*Briareum* species	(95)	

(1S*,2S*,4R*,5Z,7S*,8R*,
9S*,10S*,11S*,12R*,17R*)-
2,4,9,12-Tetraacetoxy-8,17-
epoxy-11-hydroxybriara-
5,13-dien-18-one

168

Briareum species

(95)

(1S*,2S*,4R*,5Z,7S*,8R*,
9S*,10S*, 11R*,12R*,17R*)-
2,4,9,12-Tetraacetoxy-8,17-
epoxy-11-hydroxybriara-
5,13-dien-18-one

169

Briareum species

(95)

(1S,2S,5Z,7S,8R,9S,
10S,11S,12R,14S,17R)-
Briareolide B

170

Briareum species

(117)

X-ray analysis
of **170**
(absolute
configuration)
(117)

Inhibition (55%) of
inflammation in the
mouse ear assay at
a dose of 50 μg/ear
(117)

Table 3 (continued)

Structure no.	Compound Name	Source	X-Ray Study	Biological Activity
171	(1S*,2S*,5Z,7S*,8R*,9S*, 10S*,11S*, 12R*,14S*,17R*)- Briareolide A	*Briareum* species	(117)	Inhibition (71%) of inflammation in the mouse ear assay at a dose of 50 µg/ear (117)
172	(1S*,2S*,3Z,6S*,7R*,8R*, 9S*,10S*,11R*,12R*,17R*)- Ptilosarcen-12-ol	*Ptilosarcus gurneyi*	(147)	

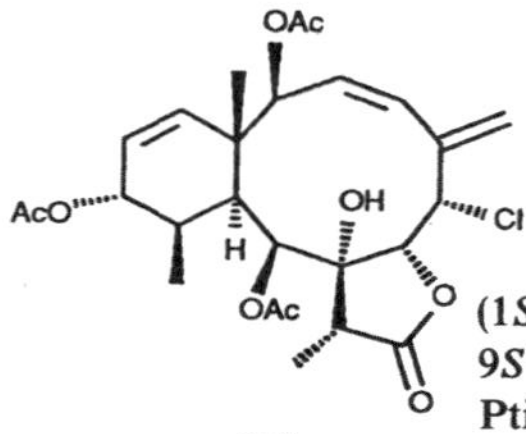

173

(1S*,2S*,3Z,6S*,7R*,8R*,9S*,10S*, 11R*,12R*,17R*)-
Ptilosarcen-12-acetate

Ptilosarcus gurneyi

(147)

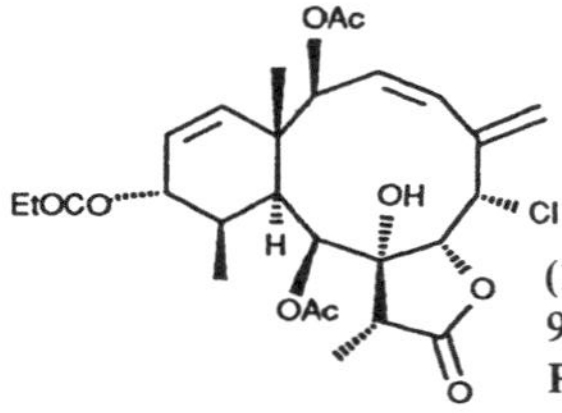

174

(1S*,2S*,3Z,6S*,7R*,8R*,9S*,10S*, 11R*,12R*,17R*)-
Ptilosarcen-12-propionate

Ptilosarcus gurneyi

(147)

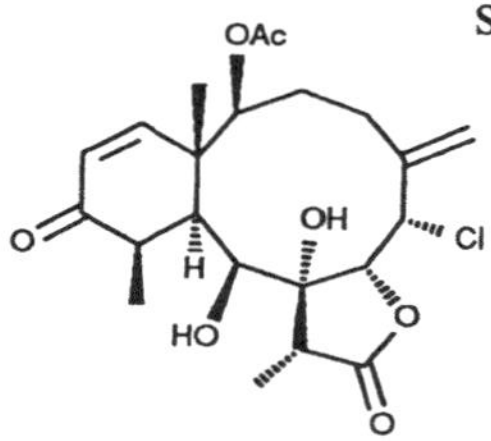

175

(1S*,2S*,6S*,7R*,8R*,9S*,10S*,11R*, 17R*)-
Solenolide E

Solenopodium species

(122)

Active against
Blowfly larvae:
ED_{100} 30–35 ppm.
Antiinflammatory
properties. Inhibitor
of the arachidonic
acid pathway
enzyme
cyclooxygenase.
Inhibitor of
Rhinovirus: IC_{50}
12.5 µg/ml
(122)

Table 3 (*continued*)

Structure no.	Compound Name	Source		X-Ray Study	Biological Activity
176	(1*S**,2*S**,3*Z*,6*S**,7*R**,8*R**,9*S**,10*S**, 11*R**,17*R**)-Ptilosarcenone	*Ptilosarcus gurneyi* *Tochuina tetraquetra*	(*147*) (*148*)	X-ray analysis of 176 (*147*)	Toxic to larvae of tobacco hornworm *Manduca sexta* at a concentration of 250 ppm (*147*)
177	(1*S**,2*S**,3*Z*,6*S**,7*R**,8*R**,9*S**,10*S**, 11*R**,17*R**)-2-Butyroxy-2-deacetoxy-ptilosarcenone	*Tochuina tetraquetra*	(*148*)		

Junceella juncea

(125)

(1R*,2S*,3Z,6S*,7R*,8R*,9S*,10S*,14S*,17R*)-Juncin B

178

Junceella fragilis

(31)

Antiviral activity against *Herpes simplex* virus I and II. Reduction in edema of 70.2%. Inhibits bee venom-derived phospholipase A_2 in *in vitro* testing (31)

(1R*,2S*,3E,6S*,7R*,8R*,9S*,10S*,14S*,17R*)-Junceellolide B

179

Briareum species

(121)

(1S*,2R*,3S*,6S*,7R*,8R*,9S*,10S*,11R*,17R*)-3,9-Diacetoxy-6-chloro-8-hydroxy-2-propanoyloxybriara-5(16),13-diene-12,18-dione

180

Table 3 (continued)

Structure no.	Compound Name	Source	X-Ray Study	Biological Activity
181	(1S*,2S*,3Z,6S*,7R*,8R*, 9S*,10S*, 11S*,12R*,17R*)- Minabein-5	Minabea species	(15)	
182	(1S*,2S*,3Z,6S*,7R*,8R*, 9S*,10S*, 11S*,12R*,17R*)- Minabein-4	Minabea species	(15)	

(1S*,2S*,6S*,7R*,8R*,9S*,
10S*, 11S*,12R*,17R*)-
Minabein-3

Minabea species

(15)

183

(1S*,2S*,3Z,6S*,7R*,8R*,
9S*,10S*,11S*,17R*)-
11-Hydroxyptilosarcenone
(Minabein-1)

Minabea species
Ptilosarcus gurneyi

(15)
(147)

184

(1S*,2S*,6S*,7R*,8R*,
9S*,10S*,11S*, 17R*)-
Minabein-2

Minabea species

(15)

185

Table 3 (*continued*)

Structure no.	Compound Name	Source	X-Ray Study	Biological Activity
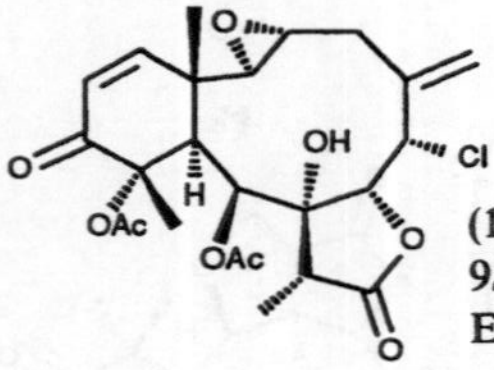 186	(1*S**,2*R**,3*R**,6*S**,7*R**,8*R**, 9*S**, 10*S**, 11*S**,17*R**)- Erythrolide C	*Erythropodium caribaeorum*	(117)	
187 R=COCH₂OCOCH₃	(1*S**,2*R**,3*R**,6*S**,7*R**,8*R**, 9*S**,10*S**, 11*S**,17*R**)- Erythrolide D	*Briareum species* *Erythropodium caribaeorum*	(117) (117)	

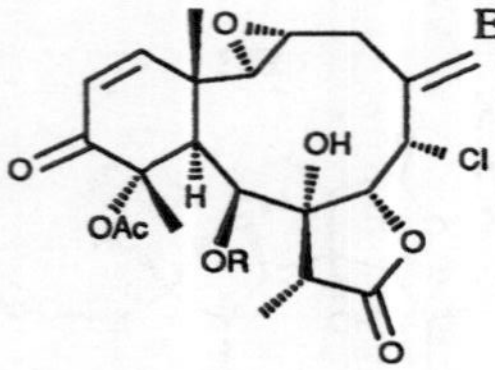

187   R=COCH₂OCOCH₃

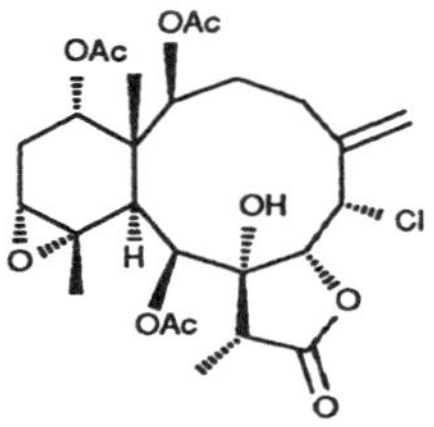

(1S*,2S*,6S*,7R*,8R*,9S*, 10S*,11S*, 12R*,14S*,17R*)- Stylatulide

Stylatula species

(99, 100)

X-ray analysis of **188** (relative configuration) (99)

Toxic to copepodite larvae of *Tisbe furcata johnsonii*: LD_{100} 0.5 ppm (99, 100)

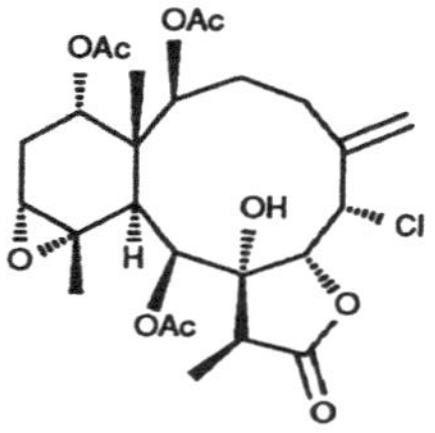

188

(1S*,2S*,6S*,7R*,8R*,9S*, 10S*,11S*, 12R*,14S*, 17S*)-17-Epi-stylatulide

Stylatula species

(100)

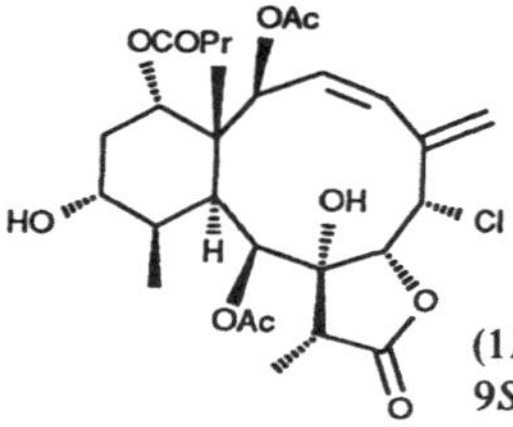

189

(1R*,2S*,3Z,6S*,7R*,8R*, 9S*,10S*,11R*, 12R*,14S*, 17R*)-Ptilosarcol

Ptilosarcus gurneyi

(147)

190

Table 3 (continued)

Structure no.	Compound Name	Source	X-Ray Study	Biological Activity
191	(1R*,2S*,3Z,6S*,7R*,8R*, 9S*,10S*, 11R*,14S*,17R*)- Ptilosarcone	Ptilosarcus gurneyi	(34, 147)	Mildly toxic to mice: LD$_{50}$ 7.4 mg/kg (34)
192	(1S,2S,3Z,6S,7R,8R,9S, 10S,11R,12S,13S,14R,17R)- Brianthein X	Briareum asbestinum Briareum polyanthes	(154, 155) (35, 153)	X-ray analysis of 192 (absolute configuration) (154, 156)

193

(1S,2S,3Z,6S,7R,8R,9S,10S,11R,12S,13S,14R,17R)-Brianthein Z

Briareum asbestinum
Briareum polyanthes

(114)
(35, 115)

Cytotoxic against P-388 leukemia cells at 10 µg/ml. Active against *Herpes simplex*-1 virus at 80 µg/ml and against mouse corona virus at 80 µg/ml (114)

194

(1S,2S,3Z,6S,7R,8R,9S,10S,11R,12S,13S,14R,17R)-Brianthein Y

Briareum asbestinum
Briareum polyanthes

(114)
(35)

Toxic to the grasshopper *Melanoplus bivittatus* (35, 111). Active against mouse corona virus at 400 µg/ml (114)

X-ray analysis of 194 (absolute configuration (35)

Table 3 (continued)

Structure no.	Compound Name	Source	X-Ray Study	Biological Activity
 195	(1S*,2S*,6S*,7R*,8R*,9S*, 10S*,11R*,12S*,13S*,14R*, 17R*)-Solenolide B	*Solenopodium species*	(122)	Active against Blowfly larvae: ED_{100} 30–35 ppm (122)
 196	(1S*,2S*,6S*,7R*,8R*,9S*, 10S*,11R*,12S*,13S*,14R*, 17R*)-Solenolide A	*Solenopodium species*	(122)	Active against Blowfly larvae: ED_{100} 30–35 ppm. Antiinflammatory properties. Inhibitor of the arachidonic acid pathway enzyme 5-lipoxygenase. Inhibitor of Rhinovirus: IC_{50} 0.39 µg/ml. Inhibitor of Polio III, Herpes, Ann Arbor and Maryland viruses (122)

197 — (1S,2S,3Z,6S,7R,8R,9S,10S,11R,12S,13S,14R,17R)- Brianthein V — *Briareum asbestinum* — (114) — X-ray analysis of **197** (absolute configuration) (114) — Cytotoxic against P-388 leukemia cells at 13 µg/ml. Active against mouse corona virus at 50 µg/ml (114)

198 — (1R*,2S*,3Z,6S*,7R*,8R*, 9S*,10S*, 11R*,14S*,17R*)- Juncin A — *Junceella juncea* — (125)

199 — (1R*,2S*,3E,6S*,7R*,8R*, 9S*,10S*, 11R*,14S*,17R*)- Junceellolide Č — *Junceella fragilis* — (31) — Reduction in edema of 80.0%. Inhibits bee venom-derived phospholipase A_2 in *in vitro* testing (31)

Table 3 (continued)

Structure no.	Compound Name	Source	X-Ray Study	Biological Activity
200	(1S,2E,4S,6S,7R,8R,9S, 10S,11S,17R)-Erythrolide **B**	*Briareum* species *Erythropodium caribaeorum*	(117) (96, 117)	
201	(1S*,2S*,3R*,6S*,7R*,8R*, 9S*,10S*, 11S*,12R*,17R*)- Erythrolide **G**	*Erythropodium caribaeorum*	(117)	

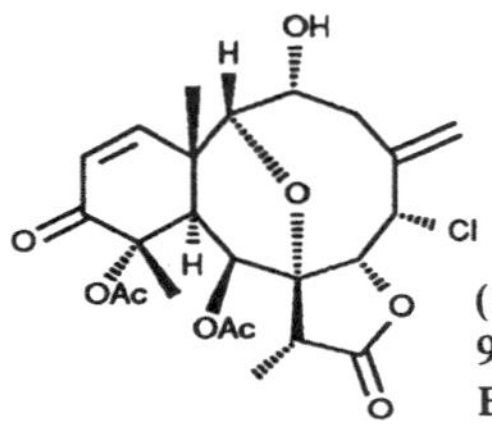

(1S*,2S*,3R*,6S*,7R*,8R*,
9S*,10S*, 11S*,17R*)-
Erythrolide E

*Erythropodium
caribaeorum*

(117)

202

(1S*,2S*,3R*,6S*,7R*,8R*,
9S*,10S*, 11S*,17R*)-
Erythrolide F

*Erythropodium
caribaeorum*

(117)

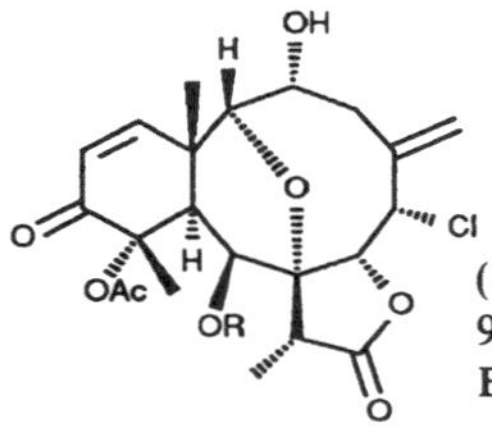

203 R=COCH₂OCOCH₃

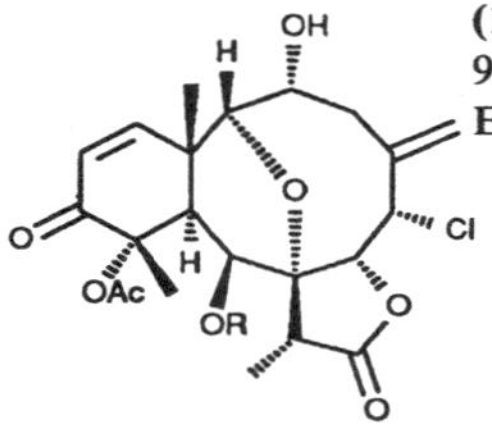

(1S*,2S*,3R*,6S*,7R*,8R*,
9S*,10S*, 11S*,17R*)-
Erythrolide I

*Erythropodium
caribaeorum*

(117)

204 R = COCH₂OH

Table 3 (continued)

Structure no.	Compound Name	Source		X-Ray Study	Biological Activity
	(1R*,2R*,3S*,4R*,6S*,7R*,8R*, 9S*, 10S*,14S*,17R*)- Junceellin	Junceella fragilis Junceella squamata	(31) (123)	X-ray analysis of 205 (123)	
205					
	(1R*,2S*,4S*,6S*,7R*,8R*, 9S*,10S*, 14S*,17R*)- Junceellolide A	Junceella fragilis	(31)		Reduction (88.4%) of edema in the mouse ear assay (31)
206					

(1R*,2S*,4S*,6S*,7R*,8R*,9S*,10S*,14S*,17S*)-14-Acetoxy-6-chloro-4,8-epoxy-9,17-dihydroxy-2-propanoyloxybriara-5(16),11-dien-18-one

Briareum species

(121)

207

(1R,2R,3S,4R,6S,7R,8R,9S,10S,11R,14S,17R)-Praelolide

Junceella fragilis
Plexaureides praelonga

(31)
(128,
129)

X-ray analysis of **208** (absolute configuration) (129)

Antiviral activity against *Herpes simplex virus* I and II (31)

208

(1R*,2S*,4S*,6S*,7R*,8R*,9S*,10S*,11R*,12R*,14S*,17R*)-Pteroidine

Pteroides laboutei

(98)

Ichthyotoxic at a concentration of 50 µg/ml (98)

209

Table 3 (*continued*)

Structure no.	Compound Name	Source	X-Ray Study	Biological Activity
210	(1R*,2S*,4S*,6S*,7R*,8R*, 9S*,10S*,11R*,12R*,14S*, 17R*)-12-O-Benzoyl-12-O-deacetylpteroidine	*Pteroides laboutei*	(98)	Ichthyotoxic at a concentration of 50 µg/ml (98)
211	(1R,2S,3Z,6S,7R,8R,9S, 10S,11S,12R,14S,17R)-Briarein A	*Briareum asbestinum*	X-ray analysis of **211** (absolute configuration) (4) (4, 32)	

Briareum asbestinum (14, 32)

(1R*,2S*,3Z,6S*,7R*,8R*,
9S*,10S*,11S*,12R*,14S*,
17R*)-Briarein B

212 R = 4xCH₃CO: C₃H₇CO

Juncella juncea (125)

(1R*,2S*,3Z,6S*,7R*,8R*,
9S*,10S*,11R*,12R*,14S*,
17R*)-Juncin D

213

Junceella juncea (125)

(1R*,2S*,3Z*,6S*,7R*,8R*,
9S*,10S*,11R*,12R*,14S*,
17R*)-Juncin C

214 R = 2xCH₃CO: iso-C₄H₉CO

Table 3 (continued)

Structure no.	Compound Name	Source	X-Ray Study	Biological Activity
215	(1S*,2S*,6S*,7R*,8R*,9S*, 10S*,11R*,12R*,14S*, 17R*)-Gemmacolide C	Junceella gemmacea	(126)	
216 R = 3xCH₃CO; iso-C₃H₇CO	(1S*,2S*,6S*,7R*,8R*, 9S*,10S*,11R*,12R*,14S*, 17R*)-Juncin F	Junceella juncea	(125)	

Z,9S*,10S*,13R*, cellin B	Junceella squamata	(124)	
4R,6S,7R,8R, ;12R,13S,14R, olide	Briareum species	(118)	X-ray analysis of **218** (absolute configuration) (118)	Reduction (37%) of edema at a concentration of 25 µg in the mouse ear assay (118)
S*,4R*,6S*,7R*,)S*,11R*,12R*, ,17R*)- C	Solenopodium species	(122)	Active against Blowfly larvae: ED_{100} 30–35 ppm (122)

Table 3 (*continued*)

Structure no.	Compound Name	Source	X-Ray Study	Biological Activity
	(1R,2R,3S,4R,6S,7R,8R, 9S,10S,11R,12R,13S,14R, 17R)-Solenolide D	*Solenopodium* species	MM2 calculations (*157*)	Active against Blowfly larvae: ED_{100} 30–35 ppm. Antiinflammatory properties. Inhibitor of Semiliki Forest- and Arbor viruses (*122*)
		(*122*)		
220				
	(1S*,2S*,3Z,6S*,7R*,8R*, 9S*,10S*,11R*,12R*,13R*; 14R*,17R*)-Gemmacolide E	*Junceella gemmacea*	(*126*)	
221				

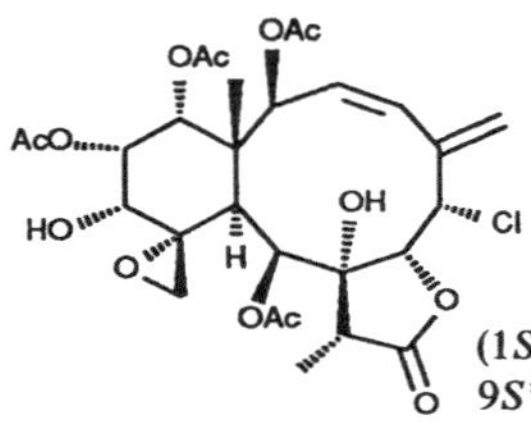

(1S*,2S*,3Z,6S*,7R*,8R*,
9S*,10S*,11R*,12R*,13R*,
14R*,17R*)-Gemmacolide D

Junceella gemmacea

(126)

222

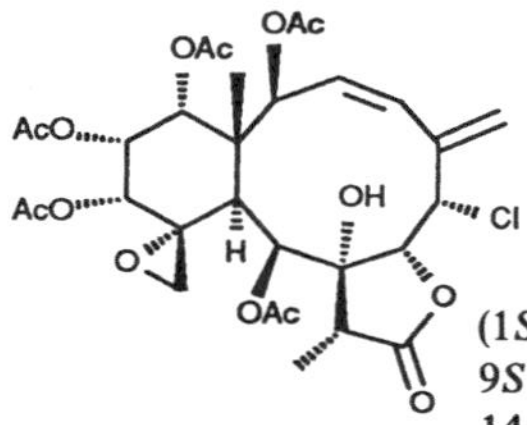

(1S*,2S*,3Z,6S*,7R*,8R*,
9S*,10S*,11R*,12R*,13R*,
14R*,17R*)-Juncin E

Junceella juncea

(125)

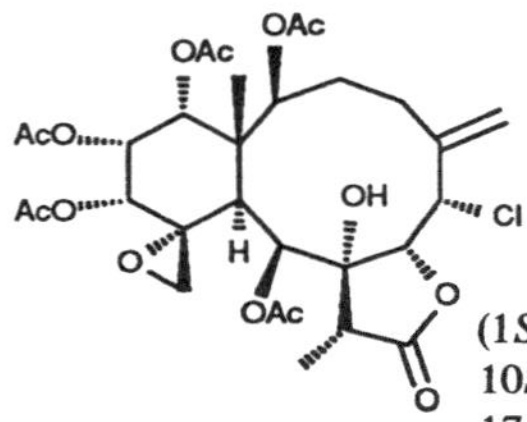

223

(1S*,2S*,6S*,7R*,8R*,9S*,
10S*,11R*,12R*,13R*,14R*,
17R*)-Gemmacolide A

Junceella gemmacea

(126)

224

Table 3 (continued)

Structure no.	Compound Name	Source	X-Ray Study	Biological Activity
	(1S*,2S*,6S*,7R*,8R*,9S*, 10S*,11R*,12R*,13R*,14R*, 17R*)-Gemmacolide B 225	Junceella gemmacea	(126)	
	3,8-Cyclized and Rearranged Cembranoid; Erythranoid			
	(1S,2E,4S,6S,7R,8R,9S, 10S,11S,13R,14R,17R)- Erythrolide A 226	Briareum species Erythropodium caribaeorum	(117) (96, 117)	X-ray analysis of 226 (absolute configuration) (96)

Cyclized and Rearranged Cembranoid; Gersolanoid

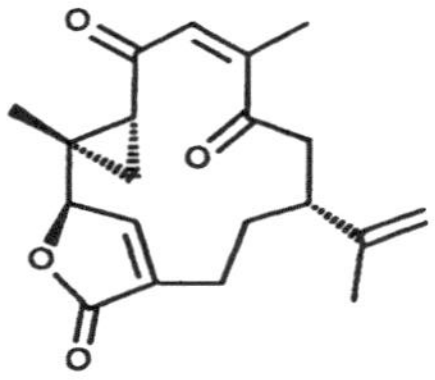

(1R*,4Z,7R*,8R*,10R*)-
Gersolide

Gersemia rubiformis (94, 141)

X-ray analysis of
227 (relative
configuration)
(94)

227

2,11:3,8-Cyclized Cembranoid

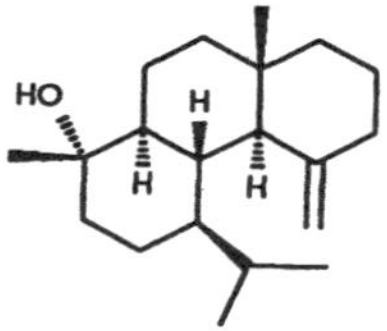

(1S,1aS,4R,4aS,5aR,8aS)-4-
Isopropyl-1,8a-dimethyl-5-
methylenetetradecahydro-
phenanthren-1-ol

Briareum species (95)

228

Addendum

The defence secretions from soldiers of the *Nasutitermes nigriceps* species collected in Peru and Mexico have been examined by VALTEROVÁ *et al.* (*158*). The results show that the Peruvian termites produce only hydroxylated trinervitanoids (**11, 12, 16, 17**) and rippertenol (**63**), while the diols **14** and **21** as well as acetylated trinervitanoids (**30, 40, 247**) are present in the Mexican species (see Tables 2 and 4). Of the latter, the triacetate **247** is a new natural product.

As many as twenty-five diterpenoids have been isolated from an Australian collection of the gorgonian *Solenopodium stechei* (*159*) (Table 5). Four of these, the solenopodins A-D (**248–251**) are classified as cladiellins. They are unique, however, since they do not possess the 2,9 ether bridge across the 10-membered carbocyclic ring.

Twenty of the new compounds, the stecholides (**166, 255–273**), are nonchlorinated briarans having an 8,17 epoxy group. One of them, named 3-acetoxy-stecholide E, appears to be identical with **166** previously isolated from a *Briareum* species (*95*) (see Table 3). The remaining diterpenoid was identified as an 11,12-epoxycembra-2,7-dien-4-ol. The occurrence of these compounds provides circumstantial evidence for the view that the cladiellans and briarans are cembrane-derived.

Three additional cladiellins (**252–254**) have been obtained from an Okinawan *Cladiella* species (*160*). They are structurally related to compound **78**, previously isolated from this species of soft coral (*133*).

Table 4. *Cyclized Cembranoid from Insects*

Structure no.	Compound Name	Source
	4,16:7,16-Cyclized Cembranoid; Trinervitanoid	
	(2R*,3R*,4S*,7R*,12S*,14S*,16S*)-Trinervita-1(15), 8(19)-diene-2,3,14-triyl triacetate	*Nasutitermes nigriceps* (158)

247

Table 5. Cyclized Cembranoids from Marine Organisms

Structure no.	Compound Name	Source	X-Ray Study	Biological Activity
	2,11-Cyclized Cembranoids; Cladiellins			
248	(1S*,2Z,6R*,7R*,9R*,10S*,11R*,14R*)-Solenopodin A	*Solenopodium stechei* (159)		
249	(1S*,2Z,6S*,7S*,9R*,10S*,11R*,14R*)-Solenopodin B	*Solenopodium stechei* (159)		
250	(1S*,2Z,6R*,7R*,10R*,11R*,14R*)-Solenopodin C	*Solenopodium stechei* (159)		

(1S*,2Z,6R*,9R*,10S*,11R*,14R*)-
Solenopodin D

Solenopodium stechei (159)

X-ray analysis of
251 (relative con-
figuration) (159)

251

(1R*,2R*,3R*,6S*,7S*,9R*,10R*,14R*)-3-
Acetoxycladiell-11-ene-6,7-diol

Cladiella species (160)

252

(1R*,2R*,3R*,6S*,7S*,9R*,10S*,11R*,
12S*,14R*)-3,6-Diacetoxy-11,12-epoxy-
cladiellan-7-ol

Cladiella species (160)

253

Table 5 (continued)

Structure no.	Compound		Source	X-Ray Study	Biological Activity
	Name				
254	(1R*,2R*,3R*,6S*,9R*,10R*,14R*)-6-Acetoxycycladiella-7(16),11-dien-3-ol		Cladiella species (160)		
3,8-Cyclized Cembranoids; Briarans 255	(1S*,2S*,4R*,5Z,7S*,8R*,9S*,10S*, 11R*,12S*,14S*,17R*)-Stecholide A		Solenopodium stechei (159)		Cytotoxic against P-388 murine leukemia cells: ED_{50} 4.5 µg/ml (159)

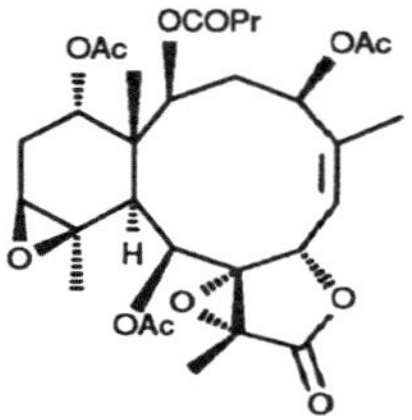

Cytotoxic against P-388 murine leukemia cells: ED$_{50}$5.4 µg/ml (159)

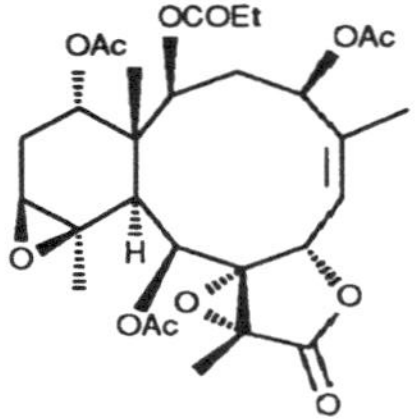

Solenopodium stechei (159)

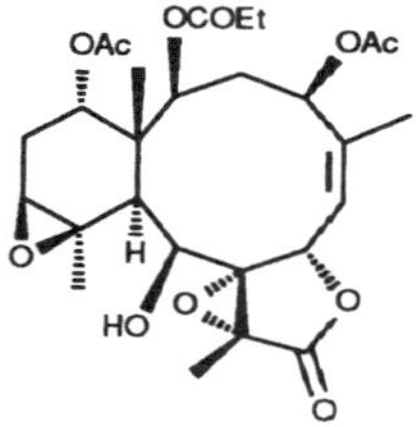

Solenopodium stechei (159)

Solenopodium stechei (159)

(1S*,2S*,4R*,5Z,7S*,8R*,9S*,10S*,11R*,12S*,14S*,17R*)-Stecholide A acetate

(1S*,2S*,4R*,5Z,7S*,8R*,9S*,10S*,11R*,12S*,14S*,17R*)-Stecholide B

(1S*,2S*,4R*,5Z,7S*,8R*,9S*,10S*,11R*,12S*,14S*,17R*)-Stecholide B acetate

256

257

258

Table 5 (continued)

Compound				
Structure no.	Name	Source	X-Ray Study	Biological Activity
259	(1S*,2S*,4R*,5Z,7S*,8R*,9S*,10S*, 11R*,12S*,14S*,17R*)-Stecholide C	Solenopodium stechei (159)		
260	(1S*,2S*,4R*,5Z,7S*,8R*,9S*,10S*, 11R*,12S*,14S*,17R*)-Stecholide C acetate	Solenopodium stechei (159)		

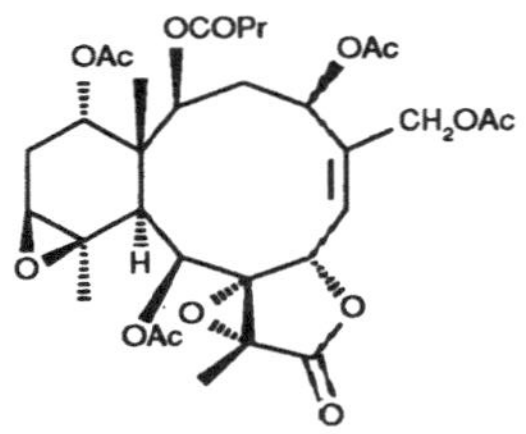

Solenopodium stechei (159)

(1S*,2S*,4R*,5Z,7S*,8R*,9S*,10S*,
11R*,12S*,14S*,17R*)-
16-Acetoxystecholide A acetate

261

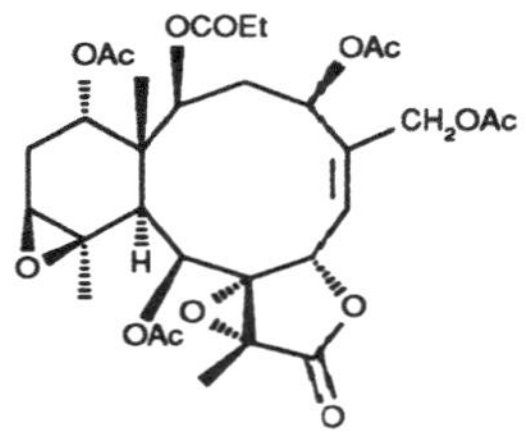

Solenopodium stechei (159)

(1S*,2S*,4R*,5Z,7S*,8R*,9S*,10S*,
11R*,12S*,14S*,17R*)-
16-Acetoxystecholide B acetate

262

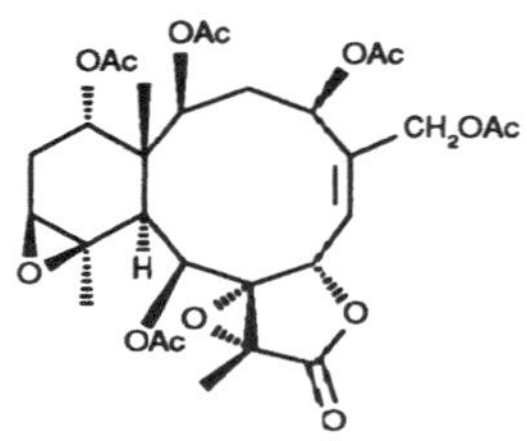

Solenopodium stechei (159)

(1S*,2S*,4R*,5Z,7S*,8R*,9S*,10S*,
11R*,12S*,14S*,17R*)-
16-Acetoxystecholide C acetate

263

Table 5 (continued)

Structure no.	Name	Source	X-Ray Study	Biological Activity
	Compound			
	(1R*,2S*,4R*,5Z,7S*,8R*,9S*,10S*, 14S*,17R*)-11,12- Deoxystecholide A acetate	Solenopodium stechei (159)		
	(1S*,2S*,4R*,5Z,7S*,8R*,9S*,10S*, 11R*,12S*,14S*,17R*)- Stecholide D	Solenopodium stechei (159)		

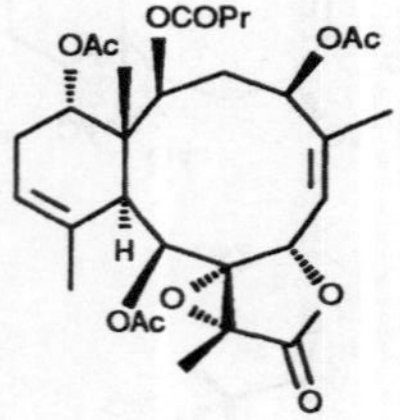

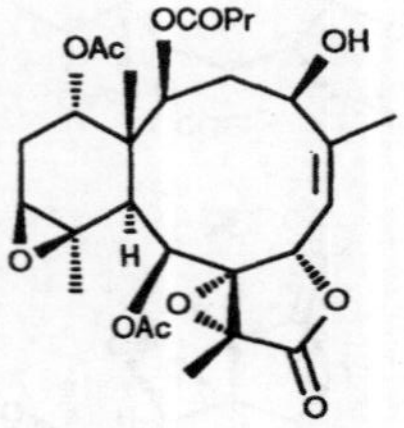

264

265

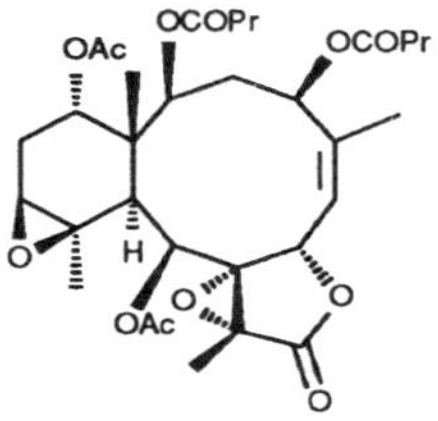

Solenopodium stechei (159)

(1S*,2S*,4R*,5Z,7S*,8R*,9S*,10S*,
11R*,12S*,14S*,17R*)-
Stecholide D butyrate

266

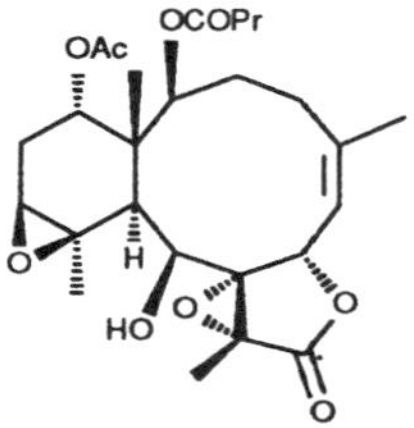

Solenopodium stechei (159)

(1S*,2S*,5Z,7S*,8R*,9S*,10S*,11R*,
12S*,14S*,17R*)-
Stecholide E

267

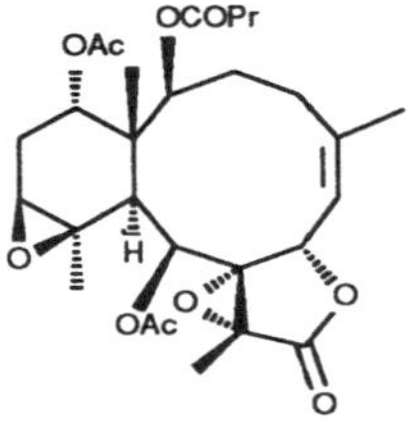

Solenopodium stechei (159)

(1S*,2S*,5Z,7S*,8R*,9S*,10S*,11R*,
12S*,14S*,17R*)-
Stecholide E acetate

268

Table 5 (continued)

Structure no.	Compound Name	Source	X-Ray Study	Biological Activity
	(1S*,2S*,5Z,7S*,8R*,9S*,10S*,11R*, 12S*,14S*,17R*)- Stecholide F	*Solenopodium stechei* (159)		
	(1S*,2S*,5Z,7S*,8R*,9S*,10S*, 14S*,17R*) 11,12-Deoxystecholide E	*Solenopodium stechei* (159)		

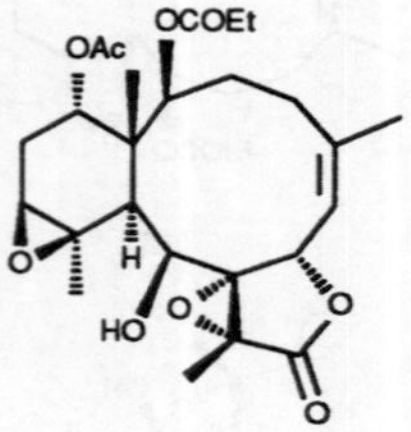

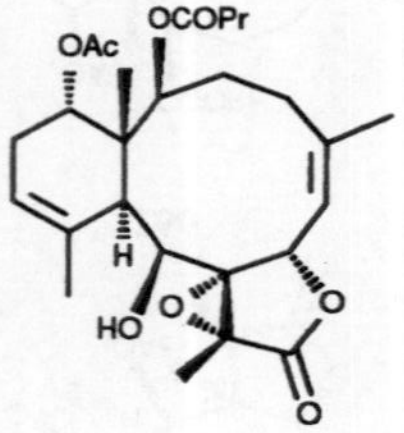

269

270

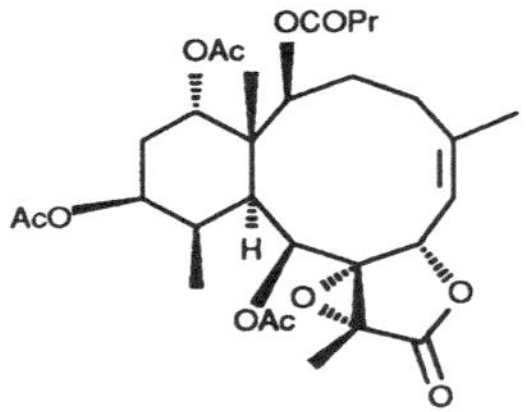

Cytotoxic against P-388 murine leukemia cells: ED$_{50}$ 10 μg/ml (159)

Solenopodium stechei (159)

Solenopodium stechei (159)

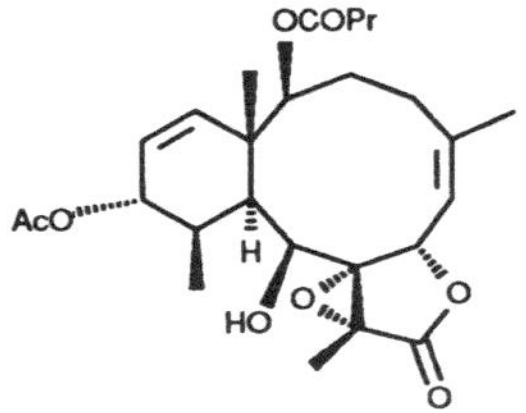

Solenopodium stechei (159)

(1S*,2S*,5Z,7S*,8R*,9S*,10S*,11R*, 12S*,14S*,17R*)-11,12-Deoxy-11-H,12-acetoxystecholide E acetate

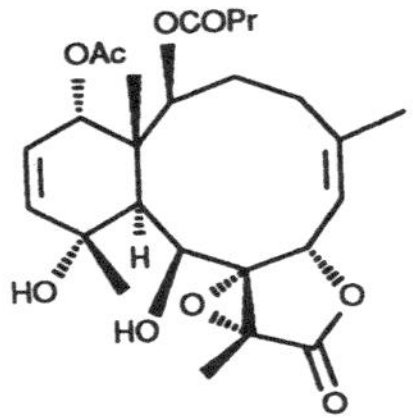

(1S*,2S*,5Z,7S*,8R*,9S*,10S*,11R*, 12R*,17R*)-Stecholide G

(1S*,2S*,5Z,7S*,8R*,9S*,10S*,11R*, 14S*,17R*)-Stecholide H

271

272

273

Acknowledgements

We are grateful to Professor Curt R. Enzell for his interest in this work, to Professor John D. Faulkner and Dr. Irena Valterová for helpful advice and to Ms. Gabriella Huss for typing this manuscript.

References

1. KENNARD, O., D.G. WATSON, L. RIVA DI SANSEVERINO, B. TURSCH, R. BOSMANS, and C. DJERASSI: Chemical Studies of Marine Invertebrates. IV. Terpenoids LXII. Eunicellin, a Diterpenoid of Gorgonian *Eunicella stricta*. X-Ray Diffraction Analysis of Eunicellin Dibromide. Tetrahedron Letters 2879–2884 (1968).

2. HYDE, R.W.: Thesis, University of Oklahoma (1966).

3. BARTHOLOMÉ, C.: Thesis, University Libre De Bruxelles (1974).

4. BURKS, J.E., D. VAN DER HELM, C.Y. CHANG, and L.S. CIERESZKO: The Crystal and Molecular Structure of Briarein A, a Diterpenoid from the Gorgonian *Briareum asbestinum*. Acta Crystallogr. **B33**, 704–709 (1977).

5. PRESTWICH, G.D., S.P. TANIS, J.P. SPRINGER, and J. CLARDY: Nasute Termite Soldier Frontal Gland Secretions. 1. Structure of Trinervi-2β,3α,9α-triol 9-*O*-Acetate, a Novel Diterpene from *Trinervitermes* Soldiers. J. Am. Chem. Soc. **98**, 6061–6062 (1976).

6. ERDTMAN, H., T. NORIN, M. SUMIMOTO, and A. MORRISON: Verticillol, a Novel Type of Conifer Diterpene. Tetrahedron Letters 3879–3886 (1964).

7. KARLSSON, B., A.-M. PILOTTI, A.-C. SÖDERHOLM, T. NORIN, S. SUNDIN, and M. SUMIMOTO: The Structure and Absolute Configuration of Verticillol, a Macrocyclic Diterpene Alcohol from the Wood of *Sciadopitys verticillata* Sieb. et Zucc. (Taxodiaceae). Tetrahedron **34**, 2349–2354 (1978).

8. BEGLEY, M.J., C.B. JACKSON, and G. PATTENDEN: Total Synthesis of Verticillene. A Biomimetic Approach to the Taxane Family of Alkaloids. Tetrahedron **46**, 4907–4924 (1990).

9. GUÉRITTE-VOEGELEIN, F., D. GUÉNARD, and P. POTIER: Taxol and Derivatives: A Biogenetic Hypothesis. J. Nat. Prod. **50**, 9–18 (1987).

10. MAHATO, S.B., B.C. PAL, T. KAWASAKI, K. MIYAHARA, O. TANAKA, and K. YAMASAKI: Structure of Cleomeolide, a Novel Diterpene Lactone from *Cleome icosandra* Linn. J. Am. Chem. Soc. **101**, 4720–4723 (1979).

11. BURKE, B.A., W.R. CHAN, V.A. HONKAN, J.F. BLOUNT, and P.S. MANCHAND: The Structure of Cleomeolide, an Unusual Bicyclic Diterpene from *Cleome viscosa* L. (Capparaceae). Tetrahedron **36**, 3489–3493 (1980).

12. PRESTWICH, G.D., R.W. JONES, and M.S. COLLINS: Terpene Biosynthesis by Nasute Termite Soldiers (Isoptera: Nasutitermitinae). Insect Biochem. **11**, 331–336 (1981).

13. FENICAL, W.: Diterpenoids. Marine Natural Products. Chemical and Biological Perspectives. (P.J. Scheuer, Ed.) New York: Academic Press **2**, 173–245 (1978).

14. STIERLE, D.B., B. CARTÉ, D.J. FAULKNER, B. TAGLE, and J. CLARDY: The Asbestinins, a Novel Class of Diterpenes from the Gorgonian *Briareum asbestinum*. J. Am. Chem. Soc. **102**, 5088–5092 (1980).

15. KSEBATI, M.B., and F.J. SCHMITZ: Diterpenes from a Soft Coral, *Minabea* sp., from Truk Lagoon. Bull. Soc. Chim. Belg. **95**, 835–851 (1986).

16. WAHLBERG, I., and A.-M. EKLUND: Cembranoids, Pseudopteranoids, and Cubitan-

oids of Natural Occurrence. Fortschr. Chem. organ. Naturstoffe **59**, 141–294 (1992).

17. RALDUGIN, V.A., and S.A. SHEVTSOV: Polycyclic Diterpenoids Biogenetically Related to Cembrenoids. Khim. Prir. Soedin. 327–342 (1987).

18. KATO, T., T. HIRUKAWA, T. UYEHARA, and Y. YAMAMOTO: The First Synthesis of (±)-3α-Acetoxy-15β-hydroxy-7,16-secotrinervita-7,11-diene, Defense Substance from a Termite Soldier. Tetrahedron Letters **28**, 1439–1442 (1987).

19. KATO, T., T. HIRUKAWA, and Y. YAMAMOTO: A Biogenetic Synthesis of (±)-Secotrinerviten-2β,3α-diol. J. Chem. Soc. Chem. Commun. 977–978 (1987).

20. HIRUKAWA, T., A. KOARAI, and T. KATO: Cyclization of Epoxyneo-cembrene Derivatives to Secotrinervitanes. J. Organ. Chem. **56**, 4520–4525 (1991).

21. DAUBEN, W.G., I. FARKAS, D.P. BRIDON, C.-P. CHUANG, and K.E. HENEGAR: Total Synthesis of (±)-Kempene-2. J. Am. Chem. Soc. **113**, 5883–5884 (1991).

22. MEHTA, G., and A.N. MURTY: Synthetic Studies toward the Novel Tetracyclic Diterpene Longipenol: Construction of the ABD Tricarbocyclic Framework. J. Organ. Chem. **55**, 3568–3572 (1990).

23. KANG, H.-J., and L.A. PAQUETTE: Claisen-Based Strategy for the de Novo Construction of Basmane Diterpenes. Enantiospecific Synthesis of (+)-7,8-Epoxy-2-basmen-6-one. J. Amer. Chem. Soc. **112**, 3252–3253 (1990).

24. PAQUETTE, L.A., and H.-J. KANG: Synthetic Studies on Basmane Diterpenes. Enantiospecific Total Synthesis of (+)-7,8-Epoxy-2-basmen-6-one by Claisen Ring Expansion. J. Amer. Chem. Soc. **113**, 2610–2621 (1991).

25. PRESTWICH, G.D. and J. VRKOČ: Standard Nomenclature of Termite Diterpenes. Sociobiology **4**, 139–140 (1979).

26. EKLUND, A.-M., J.-E. BERG, and I. WAHLBERG: Tobacco Chemistry. 73. 4,6,8-Trihydroxy-11-capnosene-2,10-dione, a New Cembrane-Derived Bicyclic Diterpenoid from Tobacco. Acta Chem. Scand. **46**, 367–371 (1992).

27. WAHLBERG, I., A.-M. EKLUND, T. NISHIDA, C.R. ENZELL, and J.-E. BERG: 7,8-Epoxy-4-basmen-6-one, a Tobacco Diterpenoid Having a Novel Skeleton. Tetrahedron Letters **24**, 843–846 (1983).

28. UEGAKI, R., T. FUJIMORI, N. UEDA, and A. OHNISHI: Cembrane-Derived Diterpenoid Having a Carbotricyclic Skeleton. Phytochem. **26**, 3029–3031 (1987).

29. RAVI, B.N., J.F. MARWOOD, and R.J. WELLS: Three New Diterpenes from the Sea Pen *Scytalium tentaculatum*. Austral. J. Chem. **33**, 2307–2316 (1980).

30. BOWDEN, B.F., J.C. COLL, W. PATALINGHUG, B.W. SKELTON, I. VASILESCU, and A.H. WHITE: Studies of Australian Soft Corals. XLII. Structure Determination of New Briaran Derivatives from *Briareum steckei* (Coelenterata, Octocorallia, Gorgonacea). Austral. J. Chem. **40**, 2085–2096 (1987).

31. SHIN, J., M. PARK, and W. FENICAL: The Junceellolides, New Anti-Inflammatory Diterpenoids of the Briarane Class from the Chinese Gorgonian *Junceella fragilis.* Tetrahedron **45**, 1633–1638 (1989).

32. SELOVER, S.J., P. CREWS, B. TAGLE, and J. CLARDY: New Diterpenes from the Common Caribbean Gorgonian *Briareum asbestinum* (Pallus). J. Organ. Chem. **46**, 964–970 (1981).

33. MORALES, J.J., D. LORENZO, and A.D. RODRIGUEZ: Application of Two-Dimensional NMR Spectroscopy in the Structural Determination of Marine Natural Products. Isolation and Total Structural Assignment of 4-Deoxyasbestinin Diterpenes from the Caribbean Gorgonian *Briareum asbestinum*. J. Nat. Prod. **54**, 1368–1382 (1991).

34. WRATTEN, S.J., W FENICAL, D.J. FAULKNER, and J.C. WEKELL: Ptilosarcone, the Toxin from the Sea Pen *Ptilosarcus gurneyi*. Tetrahedron Letters 1559–1562 (1977).

35. GRODE, S.H., T.R. JAMES, J.H. CARDELLINA II, and K.D. ONAN: Molecular Structures of the Briantheins, New Insecticidal Diterpenes from *Briareum polyanthes*. J. Organ. Chem. **48**, 5203–5207 (1983).

36. EKLUND, A.-M., J.-E. BERG, and I. WAHLBERG: To be published.

37. WAHLBERG, I., A.-M. EKLUND, I. FORSBLOM, and J.-E. BERG: To be published.

38. ROBERTS, D.L. and R.L. ROWLAND: Macrocyclic Diterpenes. α- and β-4,8,13-Duvatriene-1,3-diols from Tobacco. J. Organ. Chem. **27**, 3989–3995 (1962).

39. ROWLAND, R.L., and D.L. ROBERTS: Macrycyclic Diterpenes Isolated from Tobacco. α- and β-3,8,13-Duvatriene-1,5-diols. J. Organ Chem. **28**, 1165–1169 (1963).

40. BRAEKMAN, J.C., D. DALOZE, A. DUPONT, J. PASTEELS, B. TURSCH, J.P. DECLERCQ, G. GERMAIN, and M. VAN MEERSSCHE: Secotrinervitane, a Novel Bicyclic Diterpene Skeleton from a Termite Soldier. Tetrahedron Letters **21**, 2761–2762, (1980).

41. PRESTWICH, G.D., S.P. TANIS, F.G. PILKIEWICZ, I. MIURA, and K. NAKANISHI: Nasute Termite Soldier Frontal Gland Secretions. 2. Structures of Trinervitene Congeners from *Trinervitermes* Soldiers. J. Am. Chem. Soc. **98**, 6062–6064, (1976).

42. DUPONT, A., J.C. BRAEKMAN, D. DALOZE, J.M. PASTEELS, and B. TURSCH: Chemical Composition of the Frontal Gland Secretions from Neo-Guinean Nasute Termite Soldiers. Bull. Soc. Chim. Belg. **90**, 485–499 (1981).

43. VALTEROVÁ, I., M. BUDĚŠÍNSKÝ, F. TUREČEK, and J. VRKOČ: Minor Diterpene Components of the Defense Secretion from the Frontal Gland of Soldiers of the Species *Nasutitermes costalis* (Holmgren). Collect. Czech. Chem. Commun. **49**, 2024–2039 (1984).

44. PRESTWICH, G.D., S.G. SPANTON, J.W. LAUHER, and J. VRKOČ: Structure of 3α-Hydroxy-15-rippertene. Evidence for 1,2-Methyl Migration during Biogenesis of a Tetracyclic Diterpene in Termites. J. Am. Chem. Soc. **102**, 6825–6828 (1980).

45. PRESTWICH, G.D., M.S. TEMPESTA, and C. TURNER: Longipenol, a Novel Tetracyclic Diterpene from the Termite Soldier *Longipeditermes longipes*. Tetrahedron Letters **25**, 1531–1532 (1984).

46. PRESTWICH, G.D.: Chemical Defense by Termite Soldiers. J. Chem. Ecol. **5**, 459–480 (1979).

47. GOH, S.H., C.H. CHUAH, Y.P. THO, and G.D. PRESTWICH: Extreme Intraspecific Chemical Variability in Soldier Defense Secretions of Allopatric and Sympatric Colonies of *Longipeditermes longipes*. J. Chem. Ecol. **10**, 929–944 (1984).

48. PRESTWICH, G.D.: Defense Mechanisms of Termites. Ann. Rev. Entomol. **29**, 201–232 (1984).

49. VALTEROVÁ, I., J. KŘEČEK, and J. VRKOČ: Frontal Gland Secretion and Ecology of the Greater Antillean Termite *Nasutitermes hubbardii* (Isoptera, Termitidae). Acta Entomol. Bohemoslov. **81**, 416–425 (1984).

50. PRESTWICH, G.D.: From Tetracycles to Macrocycles. Chemical Diversity in the Defense Secretions of Nasute Termites. Tetrahedron **38**, 1911–1919 (1982).

51. PRESTWICH, G.D., S.G. SPANTON, S.H. GOH, and Y.P. THO: New Tricyclic Diterpene Propionate Esters from a Termite Soldier Defense Secretion. Tetrahedron Letters **22**, 1563–1566 (1981).

52. BIRCH, A.J., W.V. BROWN, J.E.T. CORRIE, and B.P. MOORE: Neocembrene-A, a Termite Trail Pheromone. J. Chem. Soc. Perkin I 2653–2658 (1972).

53. PRESTWICH, G.D.: Interspecific Variation of Diterpene Composition of *Cubitermes* Soldier Defense Secretions. J. Chem. Ecol. **10**, 1219–1231 (1984).

54. WIEMER, D.F., J. MEINWALD, G.D. PRESTWICH, and I. MIURA: Cembrene A and (3Z)-Cembrene A: Diterpenes from a Termite Soldier (Isoptera Termitidae Termitinae). J. Organ. Chem. **44**, 3950–3952 (1979).

55. MCDOWELL, P.G., and G.W. OLOO: Isolation, Identification, and Biological Activity

of Trail-Following Pheromone of Termite *Trinervitermes bettonianus* (Sjöstedt) (Termitidae: Nasutitermitinae). J. Chem. Ecol. **10**, 835–851 (1984).

56. PRESTWICH, G.D.: Termite Chemical Defense: New Natural Products and Chemosystematics. Sociobiology **4**, 127–138 (1979).

57. PRESTWICH, G.D.: Chemical Systematics of Termite Exocrine Secretions. Ann. Rev. Ecol. Syst. **14**, 287–311 (1983).

58. PRESTWICH, G.D.: Isolation and Identification of Diterpenes from Termite Soldier. Methods Enzymol. **110**, 417–425 (1985).

59. BAKER, R., and R.H. HERBERT: Insect Pheromones and Related Natural Products. Nat. Prod. Rep. **1**, 301–318 (1984).

60. GOH, S.H., S.L. TONG, and Y.P. THO: Gas Chromatography – Mass Spectrometry Analysis of Termite Defence Secretions in the Subfamily Nasutitermitinae. Mikrochim. Acta **1**, 219–229 (1982).

61. PRESTWICH, G.D., S.H. GOH, and Y.P. THO: Termite Soldier Chemotaxonomy. A New Diterpene from the Malaysian Nasute Termite *Bulbitermes singaporensis*. Experientia **37**, 11–13 (1981).

62. BAKER, R., and S. WALMSLEY: Soldier Defense Secretions of the South American Termites *Cortaritermes silvestri*, *Nasutitermes* sp. ND and *Nasutitermes kemneri*. Tetrahedron **38**, 1899–1910 (1982).

63. PRESTWICH, G.D.: Defence Secretion of the Black Termite, *Grallatotermes africanus* (Termitidae, Nasutitermitinae). Insect Biochem. **9**, 563–567 (1979).

64. CHUAH, C.H., S.H. GOH, G.D. PRESTWICH, and Y.P. THO: Soldier Defense Secretions of the Malaysian Termite, *Hospitalitermes umbrinus* (Isoptera, Nasutitermitinae). J. Chem. Ecol. **9**, 347–356 (1983).

65. CHUAH, C.H., S.H. GOH, and Y.P. THO: Soldier Defense Secretions of the Genus *Hospitalitermes* in Peninsular Malaysia. J. Chem. Ecol. **12**, 701–712 (1986).

66. GOH, S.H., C.H. CHUAH, J.C. BELOEIL, and N. MORELLET: High Molecular Weight Diterpenes and a New C-17 Methylated Trinervitene Skeleton from the Malaysian Termite *Hospitalitermes umbrinus*. Tetrahedron Letters **29**, 113–116 (1988).

67. GUSH, T.J., B.L. BENTLEY, G.D. PRESTWICH, and B.L. THORNE: Chemical Variation in Defensive Secretions of Four Species of *Nasutitermes*. Biochem. System. Ecol. **13**, 329–336 (1985).

68. VRKOČ, J., M. BUDĚŠÍNSKÝ, and P. SEDMERA: Structure of 2α,3α-Dihydroxy- and 2α,3β-Dihydroxy-1(15),8(19)-trinervitadienes from *Nasutitermes costalis* (Holmgren). Collect. Czech. Chem. Commun. **43**, 2478–2485 (1978).

69. PRESTWICH, G.D.: Interspecific Variation in the Defence Secretions of *Nasutitermes* Soldiers. Biochem. System. Ecol. **7**, 211–221 (1979).

70. CERRINI, S., D. LAMBA, I. VALTEROVÁ, M. BUDĚŠÍNSKÝ, J. VRKOČ, and F. TUREČEK: Methyl 3α,6α-Diacetoxy-10-oxo-(7α)-Kemp-11-en-20-oate: The Revised Structure for the Kempane Derivative from the Frontal Gland Secretion of *Nasutitermes costalis* Soldiers. Collect. Czech. Chem. Commun. **52**, 707–713 (1987).

71. VALTEROVÁ, I., S. VAŠÍČKOVÁ, M. BUDĚŠÍNSKÝ, and J. VRKOČ: Constituents of Frontal Gland Secretion of Peruvian Termites *Nasutitermes ephratae*. Collect. Czech. Chem. Commun. **51**, 2884–2895 (1986).

72. VALTEROVÁ, I., J. KŘEČEK, and J. VRKOČ: Intraspecific Variation in the Defence Secretions of *Nasutitermes ephratae* Soldiers and the Biological Activity of Some of Their Components. Biochem. System. Ecol. **17**, 327–332 (1989).

73. PRESTWICH, G.D., and M.S. COLLINS: Chemotaxonomy of *Subulitermes* and *Nasutitermes* Termite Soldier Defense Secretions. Evidence Against the Hypothesis of Diphyletic Evolution of the Nasutitermitinae. Biochem. System. Ecol. **9**, 83–88 (1981).

74. BRAEKMAN, J.C., D. DALOZE, A. DUPONT, J.M. PASTEELS, and R. OTTINGER: New Trinervitane Diterpenes from Neo-Guinean *Nasutitermes* sp. Bull. Soc. Chim. Belg. **93**, 291–297 (1984).

75. PRESTWICH, G.D., B.A. SOLHEIM, J. CLARDY, F.G. PILKIEWICZ, I. MIURA, S.P. TANIS, and K. NAKANISHI: Kempene-1 and -2, Unusual Tetracyclic Diterpenes from *Nasutitermes* Termite Soldiers. J. Am. Chem. Soc. **99**, 8082–8083 (1977).

76. BRAEKMAN, J.C., D. DALOZE, A. DUPONT, J.M. PASTEELS, P. LEFEUVE, C. BORDEREAU, J.P. DECLERCQ, and M. VAN MEERSSCHE: Chemical Composition of the Frontal Gland Secretion from Soldiers of *Nasutitermes lujae* (Termitidae, Nasutitermitinae). Tetrahedron **39**, 4237–4241 (1983).

77. VALTEROVÁ, I., M. BUDĚŠÍNSKÝ, J. VRKOČ, and G.D. PRESTWICH: (8Z)-1(15),8(9)-Trinervitadien-3α-ol from *Nasutitermes nigriceps* Termites. The Revised Structure for a Defense Compound of *Trinervitermes gratiosus* Termites. Collect. Czech. Chem. Commun. **55**, 1580–1585 (1990).

78. PRESTWICH, G.D., J.W. LAUHER, and M.S. COLLINS: Two New Tetracyclic Diterpenes from the Defense Secretion of the Neotropical Termite *Nasutitermes octopilis*. Tetrahedron Letters 3827–3830 (1979).

79. VRKOČ, J., M. BUDĚŠÍNSKÝ, and P. SEDMERA: Structure of Trinervitene Diterpenoids from *Nasutitermes rippertii* (Rambur). Collect. Czech. Chem. Commun. **43**, 1125–1133 (1978).

80. BRAEKMAN, J.C., D. DALOZE, J.M. PASTEELS, and Y. ROISIN: Two New C-20 Substituted Trinervitane Diterpenes from a Neo-Guinean *Nasutitermes* sp. Bull. Soc. Chim. Belg. **95**, 915–919 (1986).

81. WEBSTER, F.X., G.D. PRESTWICH, S.-K. PARK, L.A. DEURING, A. DALJEET, and B.L. BENTLEY: Isolation and Tentative Identification of Rojofuran, an Unstable Furanoditerpene from the Defense Secretion of a Venezuelan *Nasutitermes* species. Pr. Nauk. Inst. Chem. Org. Fiz. Politech. Wroclaw **33**, 449–452 (1988).

82. PRESTWICH, G.D.: Isotrinervi-2β-ol. Structural Isomers in the Defense Secretions of Allopatric Populations of the Termite *Trinervitermes gratiosus*. Experientia **34**, 682–684 (1978).

83. BRAEKMAN, J.C., D. DALOZE, A. DUPONT, J.M. ARRIETA, C. PICCINNI-LEOPARDI, G. GERMAIN, and M. VAN MEERSSCHE: 3α-Hydroxy-8β-trinervita-1,11-diene: a Novel Diterpene from Two *Trinervitermes* species. Bull. Soc. Chim. Belg. **92**, 111–114 (1983).

84. PRESTWICH, G.D.: Chemical Composition of the Soldier Secretions of the Termite *Trinervitermes gratiosus*. Insect Biochem. **7**, 91–94 (1977).

85. PRESTWICH, G.D., and D. CHEN: Soldier Defense Secretions of *Trinervitermes bettonianus* (Isoptera, Nasutitermitinae): Chemical Variation in Allopatric Populations. J. Chem. Ecol. **7**, 147–157 (1981).

86. BRAEKMAN, J.C., D. DALOZE, A. DUPONT, J.M. PASTEELS, and G. JOSENS: Diterpene Composition of Defense Secretion of Four West African *Trinervitermes* Soldiers. J. Chem. Ecol. **10**, 1363–1370 (1984).

87. VALTEROVÁ, I., J. KŘEČEK, and J. VRKOČ: Chemical Composition of Frontal Gland Secretion in Soldiers of *Velocitermes velox* (Isoptera, Termitidae) and Its Biological Activity. Acta Entomol. Bohemoslov. **85**, 241–248 (1988).

88. BAKER, R., A.J. ORGAN, K. PROUT, and R. JONES: Isolation of a Novel Triacetoxysecotrinervitane from the Termite *Constrictotermes cyphergaster* (Termitidae, SubFamily Nasutitermitinae). Tetrahedron Letters **25**, 579–580 (1984).

89. D'AMBROSIO, M., A. GUERRIERO, and F. PIETRA: 82. Coralloidolide F, the First Example of a 2,6-Cyclized Cembranolide: Isolation from the Mediterranean Alcyonacean Coral *Alcyonium coralloides*. Helv. Chim. Acta **73**, 804–807 (1990).

90. BOWDEN, B.F., J.C. COLL, and M.C. DAI: Studies of Australian Soft Corals. XLIII. The Structure Elucidation of a New Diterpene from *Alcyonium molle*. Austral. J. Chem. **42**, 665–673 (1989).

91. D'AMBROSIO, M., A. GUERRIERO, and F. PIETRA: 189. Sarcodictyin A and Sarcodictyin B, Novel Diterpenoidic Alcohols Esterified by (*E*)-*N*(1)-Methylurocanic Acid. Isolation from the Mediterranean Stolonifer *Sarcodictyon roseum*. Helv. Chim. Acta **70**, 2019–2027 (1987).

92. D'AMBROSIO, M., A. GUERRIERO, and F. PIETRA: 105. Isolation from the Mediterranean Stoloniferan Coral *Sarcodictyon roseum* of Sarcodictyin C, D, E and F, Novel Diterpenoidic Alcohols Esterified by (*E*)- or (*Z*)-*N*(1)-Methylurocanic Acid. Failure of the Carbon-Skeleton Type as a Classification Criterion. Helv. Chim. Acta **71**, 964–976 (1988).

93. KASHMAN, Y.: A New Diterpenoid Related to Eunicellin and Cladiellin from a *Muricella* sp. Tetrahedron Letters **21**, 879–880 (1980).

94. WILLIAMS, D.E., R.J. ANDERSEN, L. PARKANYI, and J. CLARDY: Gersolide, a Diterpenoid with a New Rearranged Carbon Skeleton from the Soft Coral *Gersemia rubiformis*. Tetrahedron Letters **28**, 5079–5080 (1987).

95. BOWDEN, B.F., J.C. COLL, and I.M. VASILESCU: Studies of Australian Soft Corals. XLVI. New Diterpenes from a *Briareum* Species (Anthozoa, Octocorallia, Gorgonacea). Austral. J. Chem. **42**, 1705–1726 (1989).

96. LOOK, S.A., W. FENICAL, D. VAN ENGEN, and J. CLARDY: Erythrolides: Unique Marine Diterpenoids Interrelated by a Naturally Occurring Di-π-methane Rearrangement. J. Am. Chem. Soc. **106**, 5026–5027 (1984).

97. D'AMBROSIO, M., A. GUERRIERO, and F. PIETRA: 176. Novel Cembranolides (Coralloidolide D and E) and a 3,7-Cyclized Cembranolide (Coralloidolide C) from the Mediterranean Coral *Alcyonium coralloides*. Helv. Chim. Acta **72**, 1590–1596 (1989).

98. CLASTRES, A., A. AHOND, C. POUPAT, P. POTIER, and S.K. KAN: Invertebres Marins du Lagon Neo-Caledonien, II. Etude Structurale de Trois Nouveaux Diterpenes Isoles du Pennatulaire *Pteroides laboutei*. J. Nat. Prod. **47**, 155–161 (1984).

99. WRATTEN, S.J.. D.J. FAULKNER, K. HIROTSU, and J. CLARDY: Stylatulide, a Sea Pen Toxin. J. Am. Chem. Soc. **99**, 2824–2825 (1977).

100. WRATTEN, S.J., and D.J. FAULKNER: Some Diterpenes from the Sea Pen *Stylatula* sp. Tetrahedron **35**, 1907–1912 (1979).

101. KEIFER, P.A., K.L. RINEHART, and I.R. HOOPER: Renillafoulins, Antifouling Diterpenes from the Sea Pansy *Renilla reniformis* (Octocorallia). J. Organ. Chem. **51**, 4450–4454 (1986).

102. FAULKNER D.J.: Marine Natural Products: Metabolites of Marine Invertebrates. Nat. Prod. Rep. **1**, 551–598 (1984).

103. FAULKNER D.J.: Marine Natural Products. Nat. Prod. Rep. **3**, 1–33 (1986).

104. FAULKNER D.J.: Marine Natural Products. Nat. Prod. Rep. **4**, 539–576 (1987).

105. FAULKNER D.J.: Marine Natural Products. Nat. Prod. Rep. **5**, 613–663 (1988).

106. FAULKNER D.J.: Marine Natural Products. Nat. Prod. Rep. **7**, 269–309 (1990).

107. FAULKNER D.J.: Marine Natural Products. Nat. Prod. Rep. **8**, 97–147 (1991).

108. TURSCH. B., J.C. BRAEKMAN, D. DALOZE, and M. KAISIN: Terpenoids from Coelenterates. Marine Natural Products. Chemical and Biological Perspectives. (P.J. Scheuer, Ed.) New York: Academic Press **2**, 247–296 (1978).

109. KREBS, H.C.: Recent Developments in the Field of Marine Natural Products with Emphasis on Biologically Active Compounds. Fortschr. Chem. organ. Naturstoffe **49**, 151–363 (1986).

110. BAKER, J.T., and R.J. WELLS: Biologically Active Substances from Australian Marine

Organisms. Natural Products as Medicinal Agents. Eds. J.L. Beal and E. Reinhard, Hippokrates Verlag, Stuttgart, 281–318 (1981).

111. CARDELLINA II, J.H.: Marine Natural Products as Leads to New Pharmaceutical and Agrochemical Agents. Pure and Applied Chem. **58**, 365–374 (1986).

112. FUSETANI, N., H. NAGATA, H. HIROTA, and T. TSUYUKI: Astrogorgiadiol and Astrogorgin, Inhibitors of Cell Division in Fertilized Starfish Eggs, from a Gorgonian *Astrogorgia* sp. Tetrahedron Letters **30**, 7079–7082 (1989).

113. OCHI, M., K. YAMADA, K. SHIRASE, H. KOTSUKI, and K. SHIBATA: Calicophirins A and B, Two New Insect Growth Inhibitory Diterpenoids from a Gorgonian Coral *Calicogorgia* sp. Heterocycles **32**, 19–21 (1991).

114. COVAL, S.J., S. CROSS, G. BERNARDINELLI, and C.W. JEFFORD: Brianthein V, a New Cytotoxic and Antiviral Diterpene Isolated from *Briareum asbestinum*. J. Nat. Prod. **51**, 981–984 (1988).

115. GRODE, S.H., T.R. JAMES, and J.H. CARDELLINA II: Brianthein Z, a New Polyfunctional Diterpene from the Gorgonian *Briareum polyanthes*. Tetrahedron Letters **24**, 691–694 (1983).

116. CARDELLINA II, J.H., T.R. JAMES, M.H.M. CHEN, and J.˙CLARDY: Structure of Brianthein W, from the Soft Coral *Briareum polyanthes*. J. Organ. Chem. **49**, 3398–3399 (1984).

117. PORDESIMO, E.O., F.J. SCHMITZ, L.S. CIERESZKO, M.B. HOSSAIN, and D. VAN DER HELM: New Briarein Diterpenes from the Caribbean Gorgonians *Erythropodium caribaeorum* and *Briareum* sp. J. Organ. Chem. **56**, 2344–2357 (1991).

118. KOBAYASHI, J., J.-F. CHENG, H. NAKAMURA, Y. OHIZUMI, Y. TOMOTAKE, T. MATSUZAKI, K.J.S. GRACE, R.S. JACOBS, Y. KATO, L.S. BRINEN, and J. CLARDY: Structure and Stereochemistry of Brianolide, a New Antiinflammatory Diterpenoid from the Okinawan Gorgonian *Briareum* sp. Experientia **47**, 501–502 (1991).

119. RALDUGIN, V.A., S.A. SHEVTSOV, and V.A. PENTEGOVA: Cyclization of Cembrane Diterpenoids. I. The Main Products of Interaction of Cembrene with Formic Àcid. Izv. Sib. Otd. Akad. Nauk SSSR, Ser. Khim. Nauk 89–94 (1985).

120. DAUBEN, W.G., J.P. HUBBELL, P. OBERHANSLI, and W.E. THIESSEN: Acid-Catalyzed Cyclization of Cembrene and Isocembrol. J. Organ. Chem. **44**, 669–673 (1979).

121. BOWDEN, B.F., J.C. COLL, I.M. VASILESCU, and P.N. ALDERSLADE: Studies of Australian Soft Corals. XLVII. New Halogenated Briaran Diterpenes from a *Briareum* species (Octocorallia, Gorgonacea). Austral. J. Chem. **42**, 1727–1734 (1989).

122. GROWEISS, A., S.A. LOOK, and W. FENICAL: Solenolides, New Antiinflammatory and Antiviral Diterpenoids from a Marine Octocoral of the Genus *Solenopodium*. J. Organ. Chem. **53**, 2401–2406 (1988).

123. YAO, J., J. QIAN, H. FAN, K. SHIH, S. HUANG, Y. LIN, and K. LONG: Crystal and Molecular Structure of Junceellin. Zhongshan Daxue Xuebao, Ziran Kexueban 83–87 (1984).

124. LONG, K., Y. LIN, and W. HUANG: Studies on the Chemical Constituents of the Chinese Gorgonia. VI. Junceellin B, a New Chlorine-Containing Diterpenoid from *Junceella squamata*. Zhongshan Daxue Xuebao, Ziran Kexueban 15–20 (1987); Chem. Abstr. **107**, 215049m (1987).

125. ISAACS, S., S. CARMELY, and Y. KASHMAN: Juncins A-F, Six New Briarane Diterpenoids from the Gorgonian *Junceella juncea*. J. Nat. Prod. **53**, 596–602 (1990).

126. HE, H.-Y., and D.J. FAULKNER: New Chlorinated Diterpenes from the Gorgonian *Junceella gemmacea*. Tetrahedron **47**, 3271–3280 (1991).

127. BOWDEN, B.F., J.C. COLL, and G.M. KÖNIG: Studies of Australian Soft Corals.

XLVIII. New Briaran Diterpenoids from the Gorgonian Coral *Junceella gemmacea.* Austral. J. Chem. **43**, 151–159 (1990).

128. LUO, Y., K. LONG, and Z. FANG: Studies on the Chemical Constituents of the Chinese Gorgonia. (III). Isolation and Identification of a New Polyacetoxy Chlorine Containing Diterpene Lactone (Praelolide). Zhongshan Daxue Xuebao, Ziran Kexueban 83–92 (1983).

129. DAI, J., Z. WAN, Z. RAO, D. LIANG, Z. FANG, Y. LUO, and K. LONG: Molecular Structure and Absolute Configuration of the Diterpene Lactone, Praelolide. Scientia Sinica Serie B, **28**, 1132–1142 (1985).

130. BOWDEN, B.F., J.C. COLL, J.M. GULBIS, M.F. MACKAY, and R.H. WILLIS: Studies of Australian Soft Corals. XXXVIII. Structure Determination of Several Diterpenes Derived from a *Cespitularia* Species (Coelenterata, Octocorallia, Xeniidae). Austral. J. Chem. **39**, 803–812 (1986).

131. KAZLAUSKAS, R., P.T. MURPHY, R.J. WELLS, and P. SCHÖNHOLZER: Two New Diterpenes Related to Eunicellin from a *Cladiella* Species (Soft Coral). Tetrahedron Letters 4643–4646 (1977).

132. HOCHLOWSKI, J.E., and D.J. FAULKNER: A Diterpene Related to Cladiellin from a Pacific Soft Coral. Tetrahedron Letters **21**, 4055–4056 (1980).

133. UCHIO, Y., M. NAKATANI, T. HASE, M. KODAMA, S. USUI, and Y. FUKAZAWA: A New Eunicellin-Based Diterpene from an Okinawan Soft Coral, *Cladiella* sp. Tetrahedron Letters **30**, 3331–3332 (1989).

134. OCHI, M., K. FUTATSUGI, H. KOTSUKI, M. ISHII, and K. SHIBATA: Litophynin A and B, Two New Insect Growth Inhibitory Diterpenoids from the Soft Coral *Litophyton* sp. Chem. Lett. 2207–2210 (1987).

135. OCHI, M., K. FUTATSUGI, Y. KUME, H. KOTSUKI, K. ASAO, and K. SHIBATA: Litophynin C, a New Insect Growth Inhibitory Diterpenoid from a Soft Coral *Litophyton* sp. Chem. Lett. 1661–1662 (1988).

136. OCHI, M., K. YAMADA, K. FUTATSUGI, H. KOTSUKI, and K. SHIBATA: Litophynin D and E, Two New Diterpenoids from a Soft Coral *Litophyton* sp. Chem. Lett. 2183–2186 (1990).

137. OCHI, M., K. YAMADA, K. FUTATSUGI, H. KOTSUKI, and K. SHIBATA: Litophynins F, G, and H, Three New Diterpenoids from a Soft Coral *Litophyton* sp. Heterocycles **32**, 29–32 (1991).

138. SHARMA, P., and M. ALAM: Sclerophytins A and B. Isolation and Structures of Novel Cytotoxic Diterpenes from the Marine Coral *Sclerophytum capitalis.* J. Chem. Soc. Perkin Trans. I. 2537–2540 (1988).

139. ALAM, M., P. SHARMA, A.S. ZEKTZER, G.E. MARTIN, X. JI, and D. VAN DER HELM: Sclerophytin C-F: Isolation and Structures of Four New Diterpenes from the Soft Coral *Sclerophytum capitalis.* J. Organ Chem. **54**, 1896–1900 (1989).

140. KUSUMI, T., H. UCHIDA, M.O. ISHITSUKA, H. YAMAMOTO, and H. KAKISAWA: Alcyonin, a New Cladiellane Diterpene from the Soft Coral *Sinularia flexibilis.* Chem. Lett. 1077–1078 (1988).

141. WILLIAMS, D.E., R.J. ANDERSEN, J.F. KINGSTON, and A.G. FALLIS: Minor Metabolites of the Cold Water Soft Coral *Gersemia rubiformis.* Can J. Chem. **66**, 2928–2934 (1988).

142. KOBAYASHI, M. and K. OSABE: Marine Terpenes and Terpenoids. VIII. Transannular Cyclization of 3,4-Epoxy-1,7,11-cembratriene Systems. Chem. Pharm. Bull. **37**, 1192–1196 (1989).

143. KOBAYASHI, M. and E. NAKANO: Stereochemical Course of the Transannular Cyclization, in Chloroform, of Epoxycembranoids Derived from the Geometrical Isomers of (14S)-14-Hydroxy-1,3,7,11-cembratetraenes. J. Organ. Chem. **55**, 1947–1951 (1990).

144. UCHIO, Y., Y. FUKAZAWA, B.F. BOWDEN, and J.C. COLL: New Diterpenes from an Australian *Pachyclavularia* species (Coelenterata, Anthozoa, Octocorallia). Tennen Yuki Kagobutsu Toronkai Koen Yoshishu **31 st.**, 548–553 (1989).

145. NATORI, T., H. KAWAI, and N. FUSETANI: Tubiporein, a Novel Diterpene from a Japanese Soft Coral *Tubipora* sp. Tetrahedron Letters **31**, 689–690 (1990).

146. CLASTRES, A., P. LABOUTE, A. AHOND, C. POUPAT, and P. POTIER: Invertebres Marins du Lagon Neo-Caledonien, III. Etude Structurale de Trois Nouveaux Diterpenes Isoles du Pennatulaire, *Cavernulina grandiflora.* J. Nat. Prod. **47**, 162–166 (1984).

147. HENDRICKSON, R.L., and J.H. CARDELLINA II: Structure and Stereochemistry of Insecticidal Diterpenes from the Sea Pen *Ptilosarcus gurneyi.* Tetrahedron **42**, 6565–6570 (1986).

148. WILLIAMS, D.E., and R.J. ANDERSEN: Terpenoid Metabolites from Skin Extracts of the Dendronotid Nudibranch *Tochuina tetraquetra.* Can. J. Chem. **65**, 2244–2247 (1987).

149. GUERRIERO, A., M. D'AMBROSIO, and F. PIETRA: 89. Verecynarmin A, a Novel Briarane Diterpenoid Isolated from Both the Mediterranean Nudibranch Mollusc *Armina maculata* and its Prey, the Pennatulacean Octocoral *Veretillum cynomorium.* Helv. Chim. Acta **70**, 984–991 (1987).

150. GUERRIERO, A., M. D'AMBROSIO, and F. PIETRA: 52. Slowly Interconverting Conformers of the Briarane Diterpenoids Verecynarmin B, C, and D, Isolated from the Nudibranch Mollusc *Armina maculata* and the Pennatulacean Octocoral *Veretillum cynomorium* of East Pyrenean Waters. Helv. Chim. Acta **71**, 472–485 (1988).

151. GUERRIERO, A., M. D'AMBROSIO, and F. PIETRA: 30. Isolation of the Cembranoid Preverecynarmin Alongside Some Briaranes, the Verecynarmins, from Both the Nudibranch Mollusc *Armina maculata* and the Octocoral *Veretillum cynomorium* of the East Pyrenean Mediterranean Sea. Helv. Chim. Acta **73**, 277–283 (1990).

152. KENNARD, O., D.G. WATSON, and L. RIVA DI SANSEVERINO: The Crystal and Molecular Structure of Eunicellin Dibromide, $C_{28}H_{42}O_9Br_2$. Acta Crystallogr. **B26**, 1038–1042 (1970).

153. BARNEKOW, D.E., and J.H. CARDELLINA II: Determining the Absolute Configuration of Hindered Secondary Alcohols – a Modified Horeau's Method. Tetrahedron Letters **30**, 3629–3632 (1989).

154. VAN DER HELM, D., R.A. LOGHRY, J.A. MATSON, and A.J. WEINHEIMER: Crystal and Molecular Structure of Brianthein X. J. Crystallogr. Spectrosc. Res. **16**, 713–720 (1986).

155. LINZ, G.S., M.J. MUSMAR, A.J. WEINHEIMER, G.E. MARTIN, and J.A. MATSON: Two-Dimensional NMR Studies of Marine Natural Products. III. Reassignment of the ^{13}C-NMR Spectrum of Brianthein-X Using Heteronuclear Relayed Coherence Transfer. Spectrosc. Lett. **19**, 545–557 (1986).

156. VAN DER HELM, D., R.A. LOGHRY, J.A. MATSON, and A.J. WEINHEIMER: ERRATUM – Crystal and Molecular Structure of Brianthein X. J. Crystallogr. Spectrosc. Res. **19**, 231 (1989).

157. CHENG, J.-F., S. YAMAMURA, Y. TERADA: Stereochemistry of the Brianolide Acetate (= Solenolide D) by the Molecular Mechanics Calculations. Tetrahedron Letters **33**, 101–104 (1992).

158. VALTEROVÁ, I., M. BUDĚŠÍNSKÝ, and J. VRKOČ: Defensive Substances from the Frontal Gland Secretion of *Nasutitermes nigriceps* Termite Soldiers. Collect. Czech. Chem. Commun. **56**, 2969–2977 (1991).

159. BLOOR, S.J., F.J. SCHMITZ, M.B. HOSSAIN, and D. VAN DER HELM: Diterpenoids from the Gorgonian *Solenopodium stechei*. J. Organ. Chem. **57**, 1205–1216 (1992).
160. UCHIO, Y., M. KODAMA, S. USUI, and Y. FUKAZAWA: Three New Eunicellin-Based Diterpenoids from an Okinawan *Cladiella* species of Soft Coral. Tetrahedron Letters **33**, 1317–1320 (1992).

(*Received March 9, 1992*)

Chemical Synthesis of Heparin Fragments and Analogues

M. Petitou[1] and C.A.A. van Boeckel[2]

[1] Sanofi Recherche, Gentilly, France
[2] Organon Scientific Development Group, Oss, The Netherlands

Contents

In this article the following abbreviations will be used:

All:	allyl
Ac:	acetyl
Bn:	benzyl
Bz:	benzoyl
Me:	methyl
Z:	benzyloxycarbonyl
tBu:	tert-butyl
MCA:	monochloroacetyl
Lev:	levulinoyl
Tf:	trifluoromethylsulphonyl
Ph:	phenyl
TBDMS:	tert-butyldimethylsilyl
THP:	tetrahydropyranyl

1. Introduction

1.1. Heparin

The history of heparin dates back to 1916 when Jay MacLean, a PhD student in Howell's laboratory, isolated an anticoagulant substance instead of the expected procoagulant phospholipids (*1, 2*). Since this substance was extracted from liver it was named heparin, in 1918, by Howell and Holt. Several years of debate then followed about the chemical nature of heparin (*3*). While it was initially thought to be a phospholipid, the polysaccharidic nature of heparin was suspected in 1925 and confirmed in the following years. However, the high complexity of the compound and the paucity of the analytical tools then available resulted in a very intricate situation where several authors proposed conflicting hypotheses regarding the nature of the different mono-saccharides present as well as their substituents. It was only in the late 1960's that the currently accepted chemical structure of heparin could be established (for a review, see (*4*)).

Heparin is a member of the glycosaminoglycan family of polysaccharides. Standard heparin preparations are mixtures of polysaccharide chains consisting in the repetition of a basic disaccharide sequence made up of a uronic acid and a glucosamine which are $1 \rightarrow 4$ linked (Fig. 1). Ten to thirty disaccharides are found in a chain, depending on its length. 2-O-Sulphate-α-L-iduronic acid (G, I) and 6-O-sulphate-N-sulphate-α-D-glucosamine (H, J) are the predominant monosaccharides and the repetition of trisulphated disaccharides like GH and IJ constitutes so-called "regular regions" which account for the major part of the structure. Other monosaccharide residues like α-L-iduronic acid (C) 6-O-sulphate-N-acetyl-α-D-glucosamine (D), β-D-glucuronic acid (E) and 3,6-di-O-sulphate-N-sulphate-α-D-glucosamine (F) are much less frequent and occur in "irregular regions". A certain number of glucosamine units are not sulphated at position 6 and 2-O-sulphate-α-D-glucuronic acid is present in some preparations. Altogether, 10 different monosaccharides (4 uronic acids and 6 glucosamines) appear in heparin, making the overall structure a very complex one (*4*).

The biosynthesis of heparin starts from a proteoglycan made up of a protein core bearing polysaccharide chains which are homopolymers of the disaccharide N-acetyl-α-D-glucosamine $(1 \rightarrow 4)$ β-D-glucuronic acid $(1 \rightarrow 4)$. These chains are then modified by an enzymatic machinery which is not yet completely understood. Different steps involve deacylation, N-sulphation, epimerisation and O-sulphation (*5*). The result of this biosynthesis is the formation of a more complex proteoglycan containing the polysaccharide chains described in the preceding paragraph, which in turn is processed by enzymes (glycosidases) to finally deliver the polysaccharide chains that can be extracted during heparin preparation.

Heparin is not only remarkable because of its very complex structure but also because of its various pharmacological properties, the most intriguing one being its anticoagulant activity (*6*). For this reason heparin became a very important drug in the prevention of venous thrombosis (undesired clotting of blood in veins) which is a frequent side

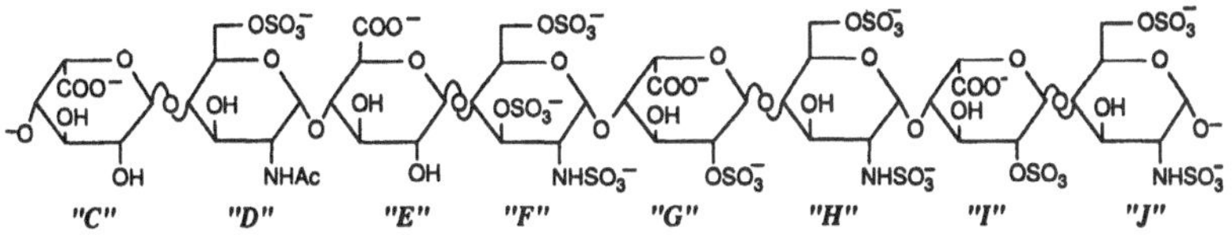

Fig. 1. Structure of a heparin fragment (see text for the meaning of "*A*", "*B*", etc. ...)

effect of surgery. The first use of heparin for this purpose dates back to 1937 but recent major improvements, like the development of low molecular weight heparin preparations which reduce the risk of side effects and allow an easier mode of administration, have contributed considerably to the spreading of this type of drug (7).

Apart from its anticoagulant activity and the associated antithrombotic activity, heparin also displays several other biological properties (8). Among these latter, its action on cell proliferation (9) on angiogenesis (10) and as antiviral (11) are particularly noteworthy.

1.2. Heparin Fragments

It is somehow paradoxical to talk about heparin fragments since we have seen in the preceding section that heparin itself is a mixture of fragments resulting from the processing of a proteoglycan by glycosidases. Indeed, the notion of heparin fragments also includes those fragments that can be obtained after cleavage of the natural polysaccharide chains by chemical or enzymic means.

In order to understand the importance of this notion of heparin fragments one has to recall some recent discoveries concerning the activity of heparin on blood clotting. On the one hand, it was firmly established in 1970–1975 that this activity is mediated by a plasma protein, antithrombin III (AT III), which is a natural blood anticoagulant and acts as a suicide-inhibitor of some blood coagulation factors, particularly factor Xa and thrombin (factor IIa), which are serine proteinases. When heparin is added to plasma it complexes with AT III (which is otherwise a weak inhibitor) so as to produce a considerable increase in the reaction rate between the serine proteinases and AT III (6). On the other hand, as we have already mentioned, heparin preparations are heterogeneous in several respects. It was thus possible, using a gel filtration system, to obtain fractions of different molecular weights displaying different anticoagulant properties, i.e. heparin chains above 5 000 Da Mw could inhibit thrombin and factor Xa in the presence of AT III, while chains with a lower molecular weight could, under similar experimental conditions, only inhibit factor Xa (12). It was also possible, in another key experiment, to fractionate heparin preparations by affinity chromatography on agarose immobilised antithrombin III. Only about one third of the heparin chains could bind to the affinity gel indicating that some heparin molecules possessed structural peculiarities responsible for this binding. These chains were only active on blood coagulation, whereas the flow-through fractions were inactive (12, 13, 14).

From these experiments it was clear that some heparin fragments were endowed with special properties, most probably related to their structure, whereupon several groups launched research programs aiming at the characterization of these fragments. The results culminated in the discovery in heparin of a unique pentasaccharide sequence (Fig. 2), which is responsible for the binding and activation of antithrombin III (*15, 16*).

It is worthy of note that this sequence contains three monosaccharide moieties which only occur to a very low extent in heparin: a 6-*O*-sulphate-*N*-acetyl-α-D-glucosamine (D); a β-D-glucuronic acid (E) and a 3,6-di-*O*-sulphate-*N*-sulphate-α-D-glucosamine (F), the latter being particularly remarkable because of the very uncommon 3-*O*-sulphation (*17*).

All these results were obtained through studies conducted on heparin fragments prepared either by chemical or by enzymatic degradation of heparin. For this reason such fragments were only available in minute amounts and several studies had to be carried out on radiolabelled material. The finding that a short particular oligosaccharide sequence could be responsible for the biological activity of a large polysaccharide molecule was a breakthrough in the field. Since highly complex molecules like heparin or heparin-like compounds (more generally glycosaminoglycans) are widely distributed in living organisms, one may speculate on the role and the ubiquity of unique structures in these polysaccharides that are associated with precise biological functions.

A major consequence of this new conception was the introduction of chemical synthesis into the field of heparin. For a long time, the lack of well defined synthetic targets had discouraged organic chemists who attempted to obtain synthetic heparin oligosaccharides. But the discovery of biologically active sequences suddenly triggered much research in this field. Due to the complexity of the structures involved the synthesis of heparin fragments became a major challenge for carbohydrate chemistry. Moreover, well defined synthetic heparin fragments are required to investigate, at a molecular level, the action of heparin on coagulation as well as on other biological systems which could be controlled by heparin or heparin-like molecules.

2. Synthesis of the Antithrombin Binding Site

The identification of a specific sequence for activation of antithrombin triggered much interest in heparin synthesis. For various reasons chemists paid little attention to this class of compounds so that only seven reports had been published between 1972 and 1976 which described synthetic approaches towards heparin fragments (*18–24*). In this section we will describe first the different syntheses that immediately followed the publication of the natural sequence and, after that, more recent work which is also related to the synthesis of the antithrombin binding site.

2.1. Strategy

The synthesis of heparin fragments is a classical problem of oligosaccharide synthesis complicated by the presence, in the target molecules, of different functional groups in specific positions. Thus hydroxyl, carboxylate, O-sulphate and N-acetyl groups are present on pentasaccharide (1) (Fig. 2). In order to introduce all of them in the proper position the first part of the synthesis consists in the preparation of a fully protected pentasaccharide with various protecting groups. In the second step these groups are sequentially removed while the definitive substituents are introduced.

Obviously the choice of the protective groups may vary according to the strategy. However, in all syntheses reported so far benzyl ethers and acetyl esters have been used, respectively, for permanent and for semi-

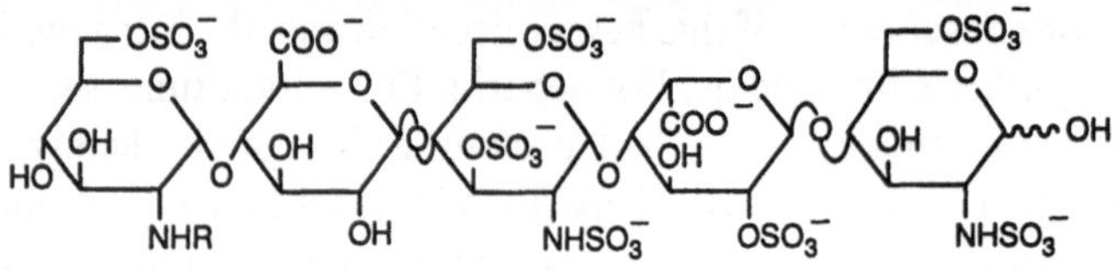

(1) R=Ac

(2) R=SO3$^-$

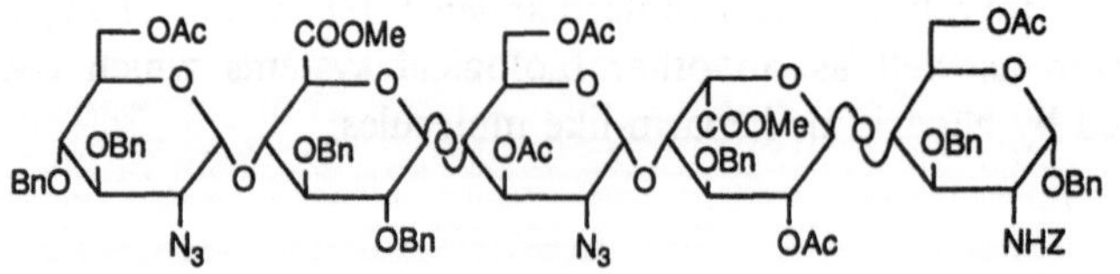

(3)

"D"   "E"   "F"   "G"   "H"

Fig. 2. Structure of the pentasaccharide sequence responsible for the binding of heparin to Antithrombin III. Note that the D unit can be either N-acetylated or N-sulphated (see Sect. 2.7.2). Compound 3 is a fully protected precursor of 2. For simplification the five monosaccharide units will be designated by D, E, F, G, H, as indicated permanent protection of hydroxyl groups. For temporary protection, several other groups were employed. Thus, a fully protected equivalent of pentasaccharide (2) is (3) (Fig. 2), in which acetyl groups replace O-sulphate, benzyl groups protect hydroxyl, azido and benzyloxycarbonyl groups replace N-sulphate and methoxycarbonyl replace carboxylate. The nature of the protecting groups is dictated by current knowledge in the field of oligosaccharide synthesis, as, for example, use of the non-participating azido group to establish 1,2-cis glycosidic linkages at the level of amino sugars or use of participating acyl groups for creating 1,2-trans interglycosidic bonds (25, 26).

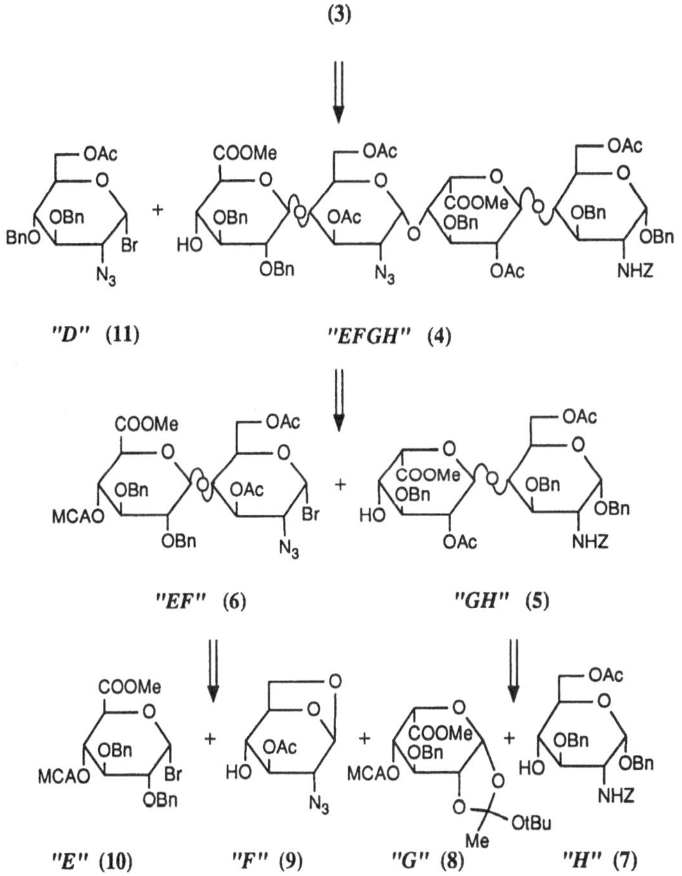

(3)

"D" (11)   "EFGH" (4)

"EF" (6)   "GH" (5)

"E" (10)   "F" (9)   "G" (8)   "H" (7)

Fig. 3. Outline of the first devised synthetic route from monosaccharides to penta-saccharide 3

Several approaches to (3) or to closely related structures have been reported. All of these use a blockwise synthesis in which (3) is obtained after coupling of two disaccharides and final addition of the non-reducing end monosaccharide (Fig. 3). The major differences between these syntheses rely on the different ways to elaborate the building blocks from various starting materials.

The problems met in the synthesis of L-iduronic and D-glucuronic acid containing building blocks of particular interest have been solved in different ways which will be reviewed first.

2.2. Preparation of D-Glucuronic Acid Derivatives

Among the many glucuronic acid derivatives described in the literature, only a few are suitable for heparin synthesis since they have to be inserted between two glucosamine moieties and therefore have to be specifically glycosylated at the 4-position and/or activated at the anomeric center to become glycosylating agents.

Within the scope of heparin synthesis, such derivatives have been obtained from D-glucose rather than from the cheap and abundant glucuronic acid, although recent attempts have been made which use D-glucuronolactone (27) as starting materials. In the first approach SINAY, PETITOU et al. (28, 29) conveniently prepared the glycosylating agent (10) from D-glucose, (Fig. 4). Oxidation at C-6 of the glucose derivative (13) was conveniently performed using JONES reagent (30) and gave (12) which was converted into (10) by isomerisation of the allyl group with pot-

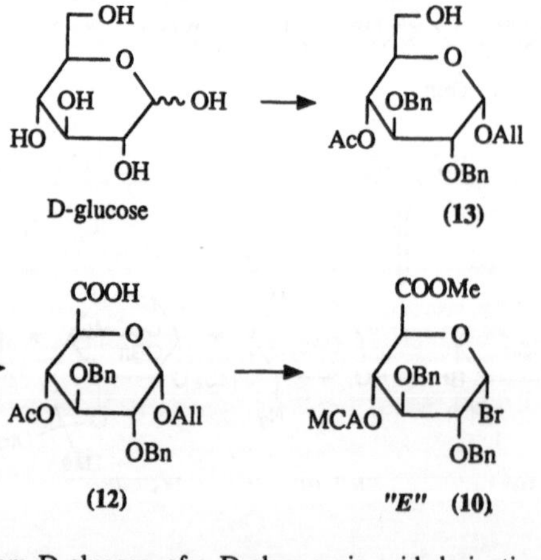

Fig. 4. Preparation, from D-glucose, of a D-glucuronic acid derivative (10) suitable for glycosylation

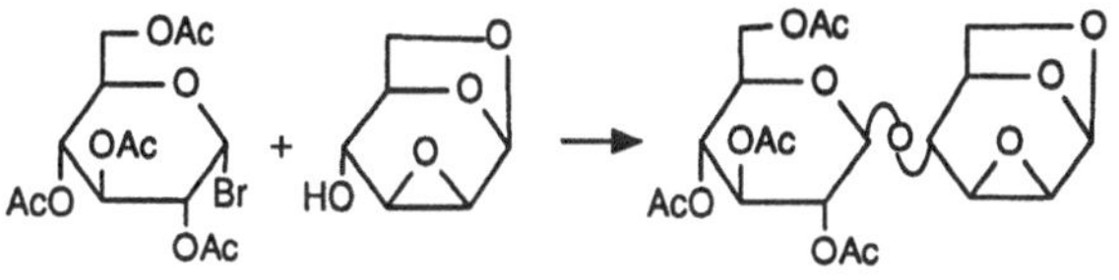

(18)   (17)   (16)

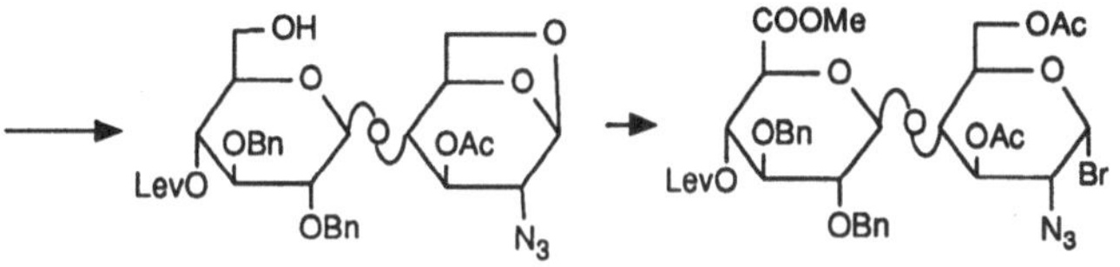

(15)   "EF"   (14)

Fig. 5. Oxidation of the glucose containing disaccharide (15) provides the glucuronic acid containing building block "EF" (14)

assium tert-butoxide, treatment with diazomethane, chloracetylation, removal of the propenyl glycoside and reaction with N,N-dimethylbromoformiminium bromide. The overall yield from 13 was 45%. In other instances air or oxygen have been employed in the presence of a platinum catalyst to selectively oxidize the primary alcohol function (31). More recently a combination of SWERN and JONES oxidation methods was used for oxidation of galactose derivatives to galacturonates (32, 33).

In other approaches towards heparin fragments oxidation has been performed at the disaccharide level. Thus van BOECKEL et al. (34) first synthesized the disaccharide (16) from acetobromoglucose (18) and the epoxide (17). Deacetylation, appropriate sequential use of protective groups, and opening of the epoxide by azide ions easily led to (15) which could be oxidized further using JONES reagent as mentioned above to give the glucuronic acid containing building block (14) (Fig. 5).

Oxidation was also performed at the disaccharide level in a synthesis described by ICHIKAWA et al. who started from the natural disaccharide cellobiose. This work will be detailed in Sect. 2.5.1.

2.3. Preparation of L-Iduronic Acid Derivatives

The preparation of L-iduronic acid derivatives is much more complicated than that of glucuronic acid derivatives since neither L-iduronic acid nor L-idose (which is a rare sugar) are available as starting materials.

Several routes for the preparation of L-idose from D-glucose (which only differ by configuration at C-5) have been proposed involving epimerisation at C-5 of glucose derivatives. This can be achieved through displacement of a sulfonate by a nucleophile (*35–37*) or through acid hydrolysis of 5,6-anhydro intermediates formed by intramolecular displacement of a leaving group at C-5 (*38, 39*).

The latter route was followed by van Boeckel et al. (*34*) in their synthesis of heparin fragments shown in (Fig. 6). The glycosylating agent (**21**) was then coupled to the glucosamine derivative (**22**) and oxidation was performed at the 6-position of (**24**) to give after deprotection (**25**), which constitutes the GH building block for (**1**) or (**2**).

Oxidation can also be effected before epimerisation (Fig. 7). Thus in the first synthesis of the antithrombin binding site (*28, 29*) the iduronic

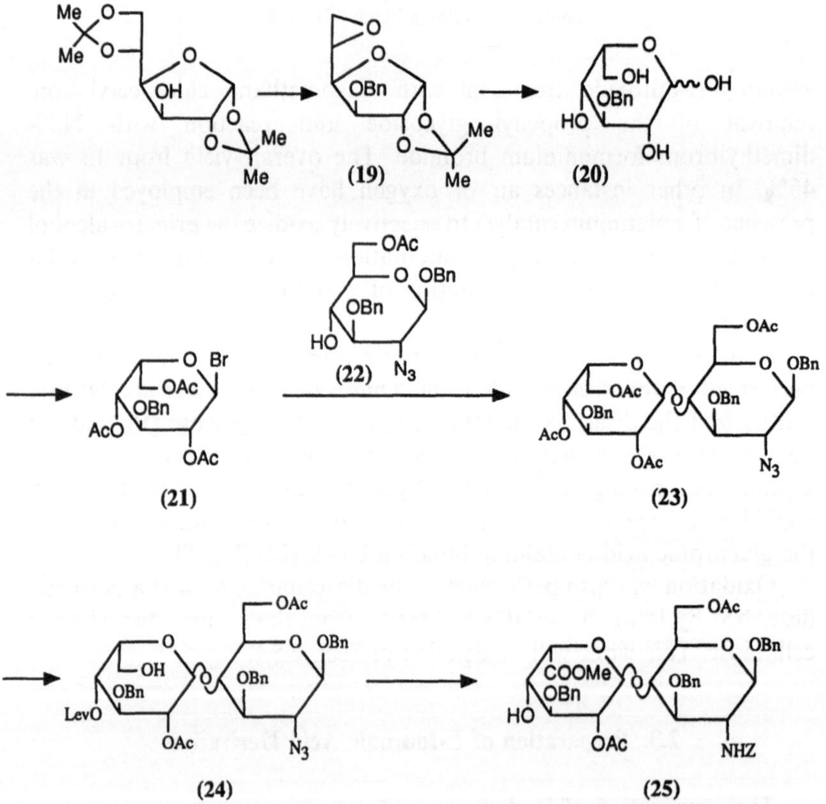

Fig. 6. Preparation of the GH building block by oxidation of disaccharide (**24**). The L-*ido* configuration was obtained from 1,2:5,6-di-*O*-isopropylidene glucose

acid ortho-ester (8) could be obtained from 1,2:5,6-di-isopropylidene α-D-glucofuranose after several steps including the inversion of configuration at C-5 of the glucuronic acid derivative (26). The use of a trifluoromethanesulfonyl leaving group in (27) in combination with trifluoroacetate as the nucleophile to give a very labile ester, that can easily be cleaved by methanolysis, is particularly important for a successful conversion since basic saponification of esters at C-5 of uronic acid can easily result in elimination. We also found that formate ion could be used as nucleophile and that mild saponification of the formate could be realised by hydrogen carbonate treatment (PETITOU et al. (77)). Compound (28) was then converted to (8) through acid hydrolysis, acetylation, ortho-ester formation, removal of acetate at the 4-position by potassium carbonate in methanol and, finally, chloroacetylation. Ortho-ester protection in (8) provides simultaneously activation of the anomeric center with respect to glycosylation, and protection of the 2-position by an acetate as required in the strategy adopted.

Several other methods for the preparation of L-idose have been tried. They invariably lead to a mixture of epimers that must be separated afterwards. Among them hydroboration of a 5,6-enoside (40–42) may be particularly useful as a route to idose and iduronic acid analogues in which conformational preferences may favour the production of the *ido* epimer.

Other epimerisation methods have been developed to obtain L-iduronic acid directly from D-glucuronic acid. BAGGETT and SMITHSON

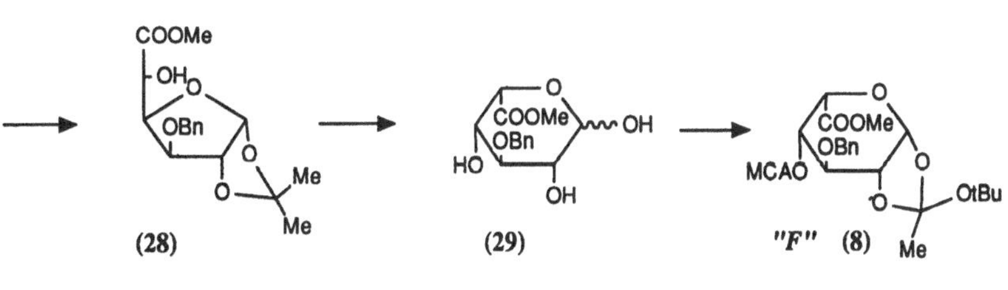

Fig. 7. Synthesis of the iduronic acid ortho-ester (8) from 1,2:5,6-di-O-isopropylidene glucose

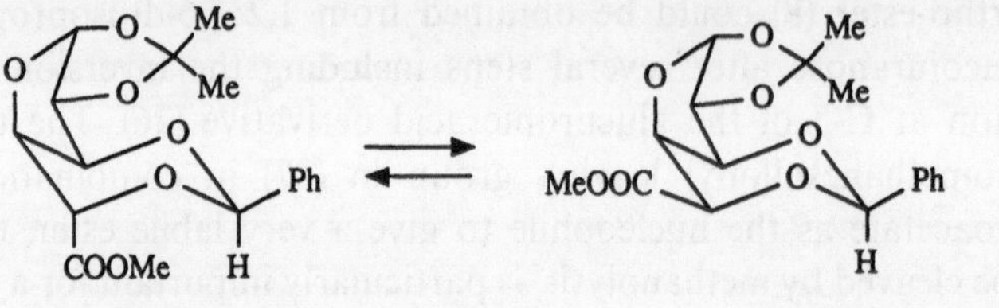

Fig. 8. Epimerisation of (30) yields the thermodynamically favored (31)

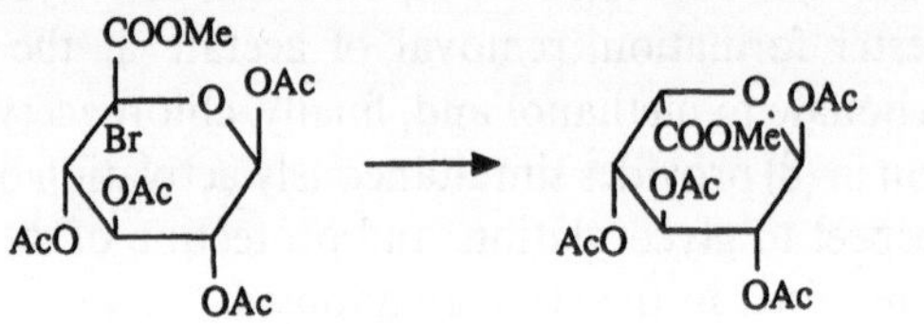

Fig. 9. Preparation of the iduronic acid derivatives (33) by reduction of (32) with tributyltin hydride

(43) epimerised methyl 3,5-O-benzylidene-1,2-isopropylidene-α-D-gluco-furanuronate (30) to the corresponding idofuranuronate (31) in which the methoxycarbonyl group is in the thermodynamically favoured equatorial orientation (Fig. 8).

A much better yield has been obtained by CHIBA and SINAY (44) who used a radical reaction to prepare 33 from 32 (Fig. 9). Unfortunately, here also a mixture of L-ido and D-gluco products is formed.

Another original approach starting from myo-inositol and involving a BAEYER-VILLIGER oxidation has also been reported (45).

2.4. Synthesis of the Fully Protected Pentasaccharides

We have already mentioned that two types of heparin binding sites (Fig. 2) may be present in the natural polysaccharide. In the first syntheses of the antithrombin binding site (28, 29, 34, 46), the authors decided to prepare the tri-N-sulphated structure (2) rather than its mono-N-acetylated counterpart (1). Due to this simplification no differentiation was required between the three glucosamine units. The more complicated synthesis of the methyl glycoside of (1) has been reported only recently by DUCHAUSSOY et al. (see Sect. 2.7).

As discussed earlier, compound (3) of Fig. 2 is a possible precursor of (2). The fully protected pentasaccharide (3) was obtained (28, 29) as depicted in (Fig. 3). Condensation of ortho-ester (8) with the glucosamine derivative (7) (unit H), followed by removal of the monochloroacetyl group gave the GH disaccharide building blocks (5) (as a matter of fact, up to now this is the only example of glycosylation using iduronic acid as glycosylating agent).

Condensation of the brominated glucuronic acid (10) (unit E) with the glucosamine precursor (9) (unit F), in dichloromethane in the presence of silver carbonate, gave a β-linked EF disaccharide (70% from (10)) together with the α-anomer (6% from (10)). Compound (6) was then easily obtained after acetolysis and bromination with titanium tetra-bromide (28, 29).

In the procedure developed by van BOECKEL et al. (34) synthons (25) (equivalent to (5)) and (14) (equivalent to (6)), the latter containing a levulinoyl instead of a monochloroacetyl group, were prepared as GH and EF disaccharide units, respectively, and subsequently oxidized to the corresponding uronic acids (Figs. 5 and 6).

In agreement with PAULSEN'S (25) studies concerning the use of 2-azido carbohydrate derivatives for the preparation of α-linked gly-cosamines, coupling of (5) and (6) in dichloromethane at − 20 °C in the presence of silver triflate and sym-collidine yielded the expected tetras-accharide in 55% yield. Coupling of (25) and (13) under the same conditions but without base gave a better yield (68%) of the trisaccharide (34). According to van BOECKEL et al., nucleophilic attack of the base at the sulphur atom of the triflate intermediate formed during the reaction may explain the detrimental effect of collidine. No β-linked product could be detected in the reaction mixture. The tetrasaccharide (4) was then obtained after treatment with thiourea in a mixture of ethanol and pyridine and coupled with the glucosamine derivative (11) (unit D), again in dichloromethane in the presence of silver triflate and sym-collidine, to give (3) (70%) together with the β-anomer (14%).

2.5. Synthesis of Building Blocks from Natural Disaccharides

2.5.1. Synthesis from Cellobiose

In the syntheses reported above, EF and GH disaccharide building blocks were prepared by coupling uronic acid derivatives (or their precursors) to precursors of glucosamine units. In an alternative proce-dure ICHIKAWA et al. started from cellobiose, a disaccharide that can be obtained from cellulose by acidic or enzymatic treatment. Regioselective modifications of cellobiose were performed as outlined in Fig. 10 in order to obtain the glucuronate and iduronate containing building blocks (6) and (40) (42, 46–49). The tetra-isopropyldisiloxane-1,3-diyl group was

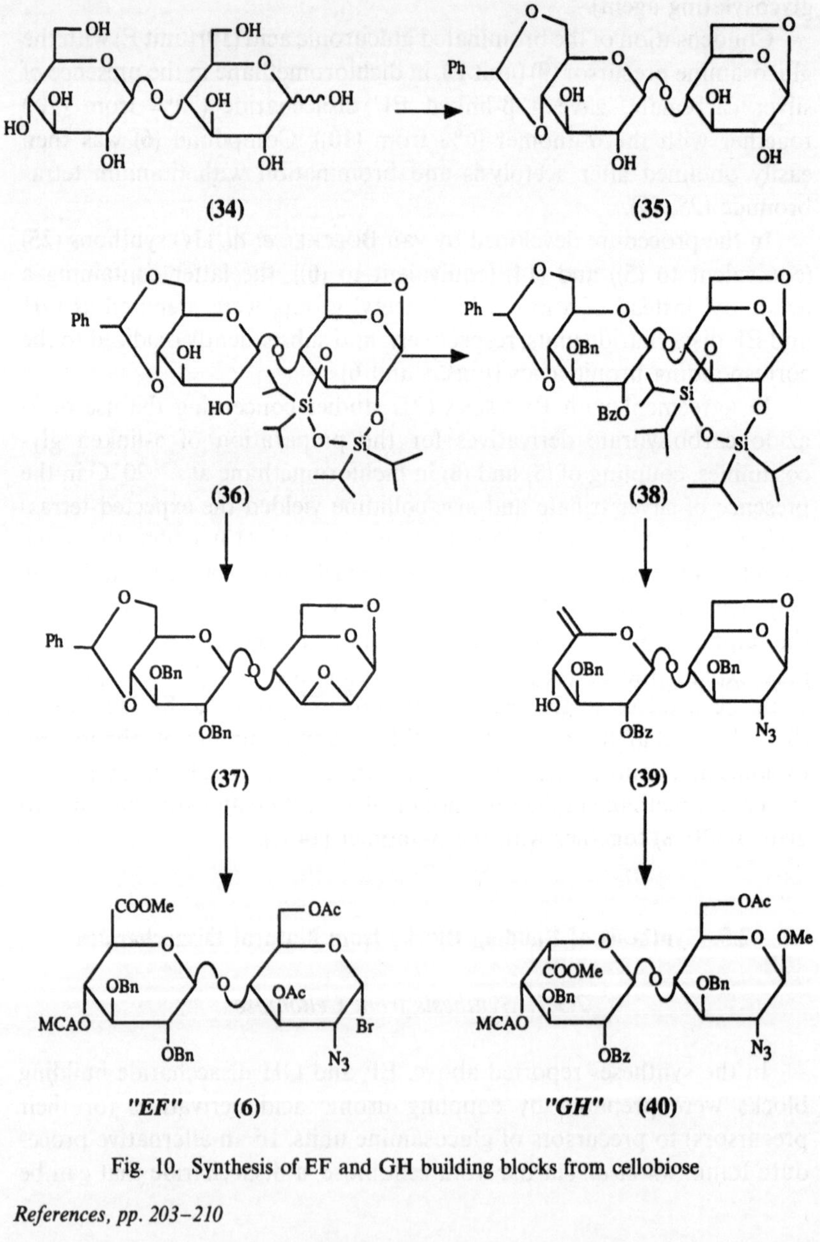

Fig. 10. Synthesis of EF and GH building blocks from cellobiose

References, pp. 203–210

used for selective protection of the 2- and 3-positions in the tetrol (**35**). EF disaccharide could then be obtained from (**36**) after benzylation, removal of the silyl ether with fluoride ions, selective tosylation at the 2-position and epoxidation to obtain (**37**). From (**37**), established methodology (Fig. 5) gave (**6**). The overall yield was 1.9% from 1,6-anhydrocellobiose.

In a modification of this process, PETITOU et al. (77) obtained the key synthon (**37**) from (**35**) after selective tosylation of the 2-position followed by basic treatment to give the epoxide and subsequent benzylation. The overall yield is comparable with the previous synthesis, but the latter process is shorter since silyl ether protection and deprotection are not required. In the same way, kinetic acetonation of 1,6-anhydrocellobiose followed by selective tosylation at the 2-position provides an efficient route to a precursor of **6** (50).

Preparation of the GH synthon (**40**) from cellobiose is even more sophisticated (42, 46, 47, 49). Thus (**35**) is first converted into the key intermediate (**36**) which after regioselective alkylation of the 3'-hydroxyl group followed by benzolyation gives (**38**). Removal of the silyl ether and epoxidation then allows introduction of azido group at the 2-position. Acid hydrolysis of the benzylidene acetal followed by tosylation and basic treatment generates the enose (**39**). Hydroboration methodology (40–42) gives the desired *ido* derivative (**40**) together with the *gluco* epimer in a 1/2 ratio. The overall yield is about 0.5% from 1,6-anhydrocellobiose.

Coupling of the EF and GH building blocks obtained by ICHIKAWA et al. was performed as already described in Sect. 2.4.

2.5.2. Synthesis from Maltose

The conversion of maltose into disaccharides having 2-amino-2 deoxy-α-D-glucose and L-idose as constituents was also studied (51, 52). The starting material is the major product formed by kinetic acetonation of maltose, namely 2,3:5,6-di-*O*-isopropylidene-4-*O*-(4,6-*O*-isopropyli-dene-α-D-glucopyranosyl)-aldehydo-D-glucose dimethyl acetal (**41**) (53). Compound (**41**) was tert-butyldimethylsilylated and converted into the α linked glucosamine containing disaccharide (**42**) via the classical sequence alcohol → ketone → imine → amine (Fig. 11) in an overall yield of 21%.

Compound (**42**) was benzylated; subsequent selective removal of the most labile isopropylidene group followed by *O*-allylation gave (**43**). Selective hydrolysis of the 5,6-isopropylidene group followed by mesylation and inversion of configuration at C-5 afforded the 5,6-di-*O*-acetyl-

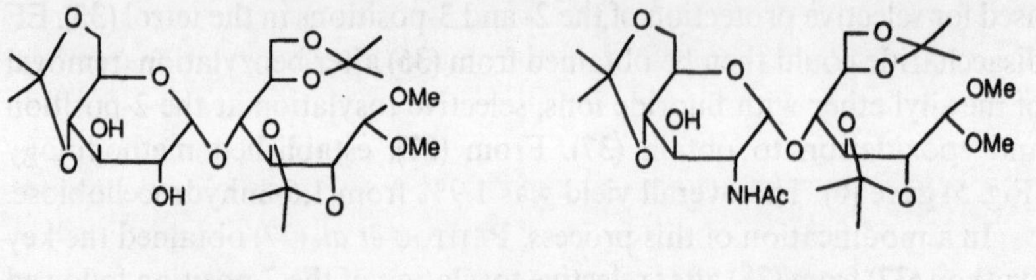

Fig. 11. The conversion of maltose into a disaccharide containing a glucosamine unit (42) involves the key synthon (41) which is the major product obtained in kinetic acetonation of maltose

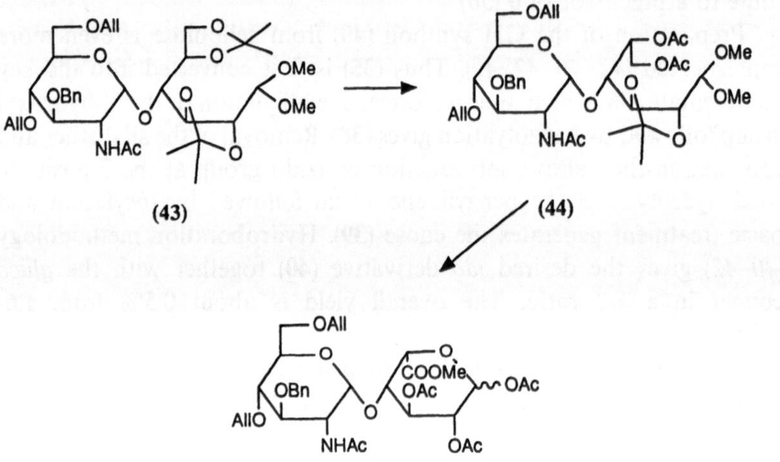

Fig. 12. Preparation of an iduronic acid containing disaccharide (45) from maltose (the disaccharide (43) was obtained from (42))

L-*ido*-product (44) which was subsequently transformed into an iduronic acid containing disaccharide, i.e. compound (45) obtained as a mixture of α and β anomeric acetates (Fig. 12) in low (probably < 1%) overall yield from (41).

2.6. Deprotection and Functionalisation

A critical point in the synthesis of heparin fragments is the conversion of the fully protected fragments to the final O,N-sulphated compounds (e.g. conversion of (46) into (47); Fig. 13). According to the strategy discussed in 2.1, the first step is saponification to liberate the hydroxyls to be sulphated. This can be carried out using sodium hydroxide in a mixture of methanol, chloroform and water to keep all the reactants in solution (29). The carboxyl groups of the uronic acid portions are also deprotected under these conditions but can be specifically methylated afterwards by diazomethane treatment. However, this extra step is not absolutely necessary since the subsequent steps of the synthesis do not affect the carboxylic acid groups. Nevertheless, the presence of the methyl esters considerably facilitates chromatographic purification on silica gel when this is necessary. The yield of the saponification step (about 70% yield) is limited due to partial cleavage of the chain by β-elimination at the 4,5 position of L-iduronic acid. The tetrasaccharide (48) isolated at

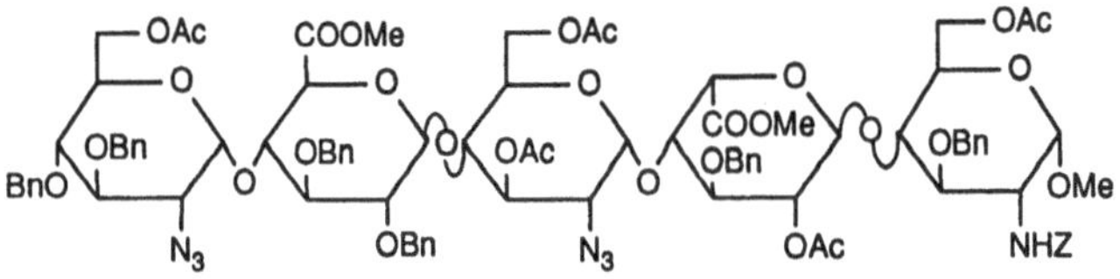

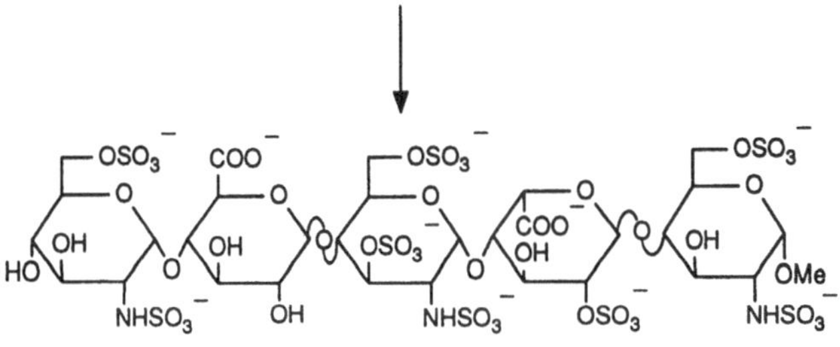

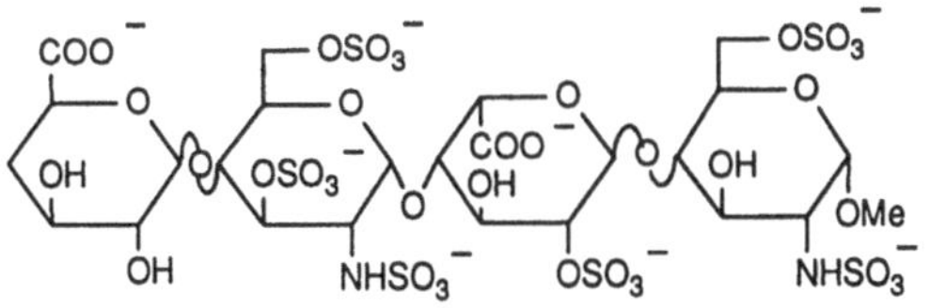

Fig. 13. The precusor (46) of the methyl glycoside of the AT III binding sequence was obtained using the strategy shown (Fig. 3) and converted into (47). The tetrasaccharide (48) was isolated as a minor side product

the end of the synthesis of (**47**) (the α-methyl analog of **2** – see below) most probably originates from this type of cleavage. Obviously the double bond has been reduced during the hydrogenolysis step.

The O-sulphation step is usually performed in dimethylformamide using a smooth sulphating agent like sulphur trioxide-trimethylamine or sulphur trioxide-triethylamine complex. The sulphated product may be isolated easily by gel filtration on a Sephadex LH20 column (29).

Hydrogenolysis of benzyl groups and concomitant reduction of the azido functions is a straightforward step although it may occasionally take a long time to reach completion. It is usually performed in a water/alcohol mixture in the presence of a palladium catalyst under a low pressure of hydrogen.

N-sulphation is performed in water with sulfur trioxide-pyridine or sulfur trioxide trimethylamine complex. The pH must be kept at around 9 which is achieved using a carbonate buffer or a pH-stat instrument (29). Here again gel filtration is the most efficient way to isolate the product which is eluted first while the lower molecular weight reagents emerge later. Further purification can be achieved by ion-exchange chromatography on Q-sepharose Fast-Flow or Mono-Q ion exchangers (76, 77).

2.7. Recent Results in the Synthesis of the Antithrombin Binding Site

2.7.1. Synthesis of the α-Methyl Glycoside of the Antithrombin Binding Site

During the hydrogenolysis step of compounds with a free anomeric end (**1** or **2**) it was found that several side products were formed. These impurities, (which were shown to behave as decasaccharides or pentadecasaccharides in gel filtration experiments (54)) originate from intermolecular reaction of free amino groups with the free reducing end obtained during the hydrogenolysis step. In order to avoid this side reaction the authors turned their attention to methyl glycosides at unit H. Indeed, deprotection and sulphation of fully protected pentasaccharides containing methyl glycosides gave much cleaner end products. Fortunately this modification does not affect the biological properties of the pentasaccharide (54).

In order to save the precious building block (**5**) a slightly different strategy was developed for the preparation of (**47**). In this case the DEF trisaccharide (**49**) (55) is condensed with GH disaccharide (**5**) to produce the fully protected pentasaccharide (**46**) in 74% yield (Fig. 14).

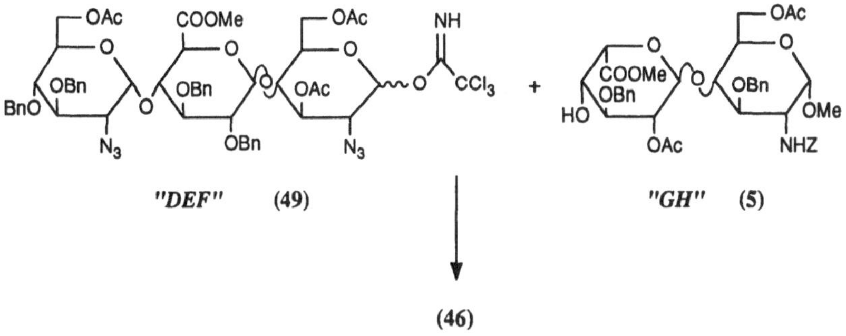

Fig. 14. Synthesis of (46) using the imidate procedure

Preparation of the α methyl pentasaccharide (47) has. been scaled up by chemists at Organon and at Sanofi and its pharmacological profile has been thoroughly investigated (56–59). As a result compound (47) is considered by the Organon and Sanofi groups as the reference compound which represents the naturally occurring pentasaccharide fragment of heparin. Biological activities of analogs are compared with that of the reference compound in an amidolytic assay which determines the AT III mediated inhibition of blood coagulation factor Xa activity. The methyl β-glycoside analogue of compound (47) has been synthesized as well and the authors reported a similar activity (49).

2.7.2. Synthesis of the N-Acetylated Antithrombin Binding Sequence

The N-acetylated (at unit D) antithrombin binding sequence represents the most widespread sequence in porcine heparin although its N-sulphated counterpart, which is characteristic of beef lung heparin, has also been reported to be present in the polysaccharide extracted from this species (60). Due to the occurrence of the N-acetylated glucosamine moiety, another strategy was adopted for its preparation (61). The key synthon for this synthesis was the trisaccharide (50): selective reduction of the single azido group followed by N-acetylation and azide addition to the epoxide provided, after acetolysis, anomeric deacetylation, and imidate formation, a DEF trisaccharide donor (51) suitable for coupling with (5). Previous attempts to couple (5) and this type of trisaccharide activated at the anomeric center in the form of a bromide had failed. Condensation of (51) with (5) followed by the usual deprotection and sulphation provided (174) (Fig. 55) which was only half as active as (47) in the biological anti-factor Xa assay (61).

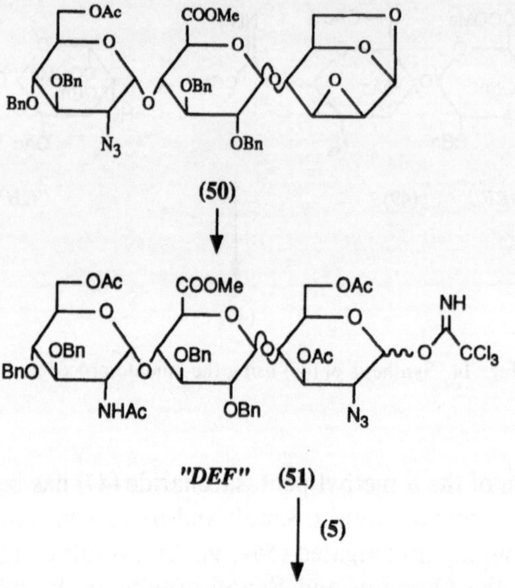

(50)

↓

"DEF"   (51)

| (5)

↓

α-methyl glycoside analog of (1) : (174).

Fig. 15. The imidate (51), N-acetylated at the D unit, is the key intermediate toward the methyl glycoside of pentasaccharide (1)

3. Synthesis and Biological Properties of Analogues of the Antithrombin Binding Site

3.1. Analogues Lacking N,O-Sulphate Groups at Defined Positions

A major question with respect to the binding properties of the heparin pentasaccharide to AT III was whether all charged groups at the pentasaccharide are required for AT III activation. The ultimate answer to this question has been provided by total synthesis of many analogues lacking charged groups at defined positions (Fig. 16) and by selective de-O-sulphation of heparin fragments. At the beginning of the eighties useful information about the role of some of the sulphate groups was obtained from studies on (bio)chemically modified heparin fragments. Lindahl et al. (62) and Atha et al. (63) established that the 6-O-sulphate group of glucosamine unit D is essential for high affinity for AT-III and for accelerating the inactivation of the coagulation factor Xa. Lindahl et al. (17) and Atha et al. (63) also pointed to a critical role for the 3-O-sulphate group at unit F; furhermore the same authors also discovered the importance of the N-sulphate groups at units F and H (63, 64). Lastly, THUNBERG et al. (16) found that the 6-O-sulphate group at unit F is non-essential, because naturally occurring heparin fragments lacking this particular 6-O-sulphate group still show high affinity towards AT III.

Since the latter studies were performed with heparin fragments larger than pentasaccharides, one had to assume that additional carbohydrate residues do not have a negative influence on the activation of AT III. Definitive proof that the unique 3-O-sulphate group at unit F is required for activation came after synthesis of the biologically inactive analogue (52) (Fig. 16) which lacks this specific charged substituent (65–67).

The synthesis (67) of analogue (52) resembles that of the α-methyl glycoside (47) of the "natural" compound (2). Since in the general strategy of the synthesis of heparin fragments (see Sect. 2.1) benzyl ethers are used for permanent protection of hydroxyl groups, the disaccharide building block (61) containing 3-O-benzyl at unit F instead of 3-O-acetyl had to be used. The route to disaccharide (61) (Fig. 17) involved coupling of bromide (58) with acceptor (59) in the presence of silver carbonate to yield the corresponding β-coupled disaccharide in 53% yield. Acetolysis of the 1,6-anhydro moiety (acetic anhydride-trifluoroacetic acid) followed by selective anomeric O-deacetylation with benzylamine gave disaccharide (60) in 85% yield. Bromination of the anomeric centre with (bromomethylene)dimethylammonium bromide gave the key bromide (61) in 48% yield. Bromide (61) could then be condensed with known building block (5), corresponding to GH, in the presence of silver triflate, to give a fully protected tetrasaccharide EFGH in 42% yield. Conversion to analogue (52) followed methodologies similar to those described in Sect. 2.6. PETITOU (65) and ATHA et al. (68) showed that analogue (52) neither binds to AT III nor induces anti-factor Xa activity.

Following a similar synthetic strategy, pentasaccharide (53) lacking the 6-O-sulphate group at H was synthesized (69). This synthesis (Fig. 18) required a suitably protected disaccharide corresponding to GH, which contains a 6-O-benzyl group instead of an 6-O-acetyl protective group at unit H (i.e. compound (65), Fig. 18).

Monosaccharide (62) was coupled with idopyranosyl bromide (21) in the presence of $Hg(CN)_2$/$HgBr_2$ to give disaccharide (63) in 55% yield. Then disaccharide (63) was deacetylated and reprotected with a 2'-O-acetyl and a 4'-O-levulinoyl ester to give (64) whose 6'-CH_2OH was oxidized under JONES conditions. After esterification with diazomethane and selective removal of the levulinoyl group by hydrazinolysis, disaccharide (65) was obtained in 17% overall yield from (63). BEETZ and VAN BOECKEL (69) used disaccharide (65) in a total synthesis along with disaccharide (14) and monosaccharide (11) to give, after standard depro-

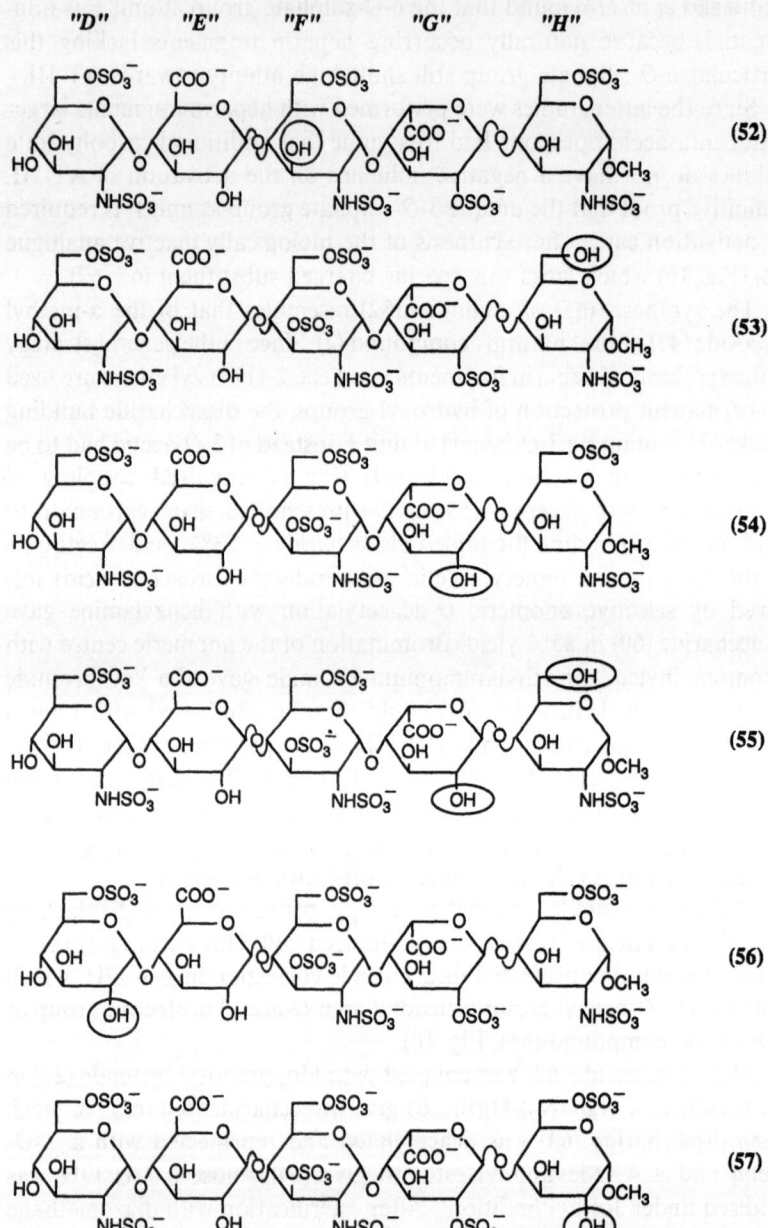

Fig. 16. Methyl glycoside analogues lacking N,O sulphate groups at defined positions

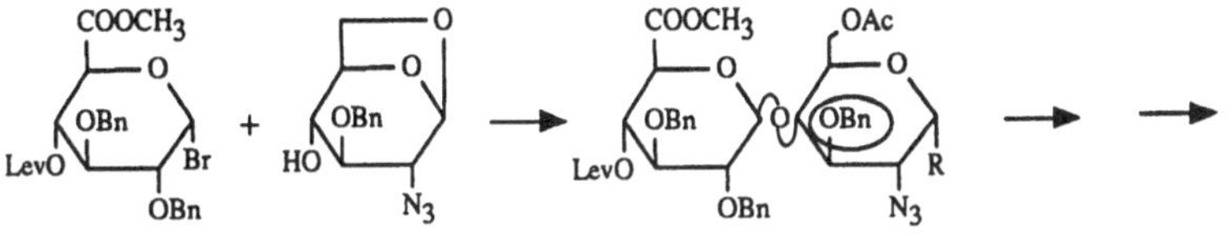

(52)

(58) (59) "*EF*" (60) R = OH

(61) R = Br

Fig. 17. Preparation of the EF building block for the synthesis of analogue (**52**), lacking 3-*O*-sulphate at F

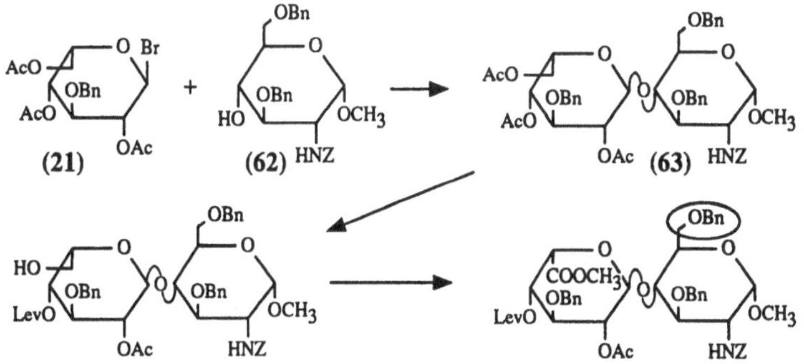

(21) (62) (63)

(64) "*GH*" (65)

Fig. 18. Preparation of the GH building block (**65**) for the synthesis of analogue (**53**), lacking 6-*O*-sulphate at H

tection-sulphation processing, analogue (**53**). Compound (**53**) displayed about 25% of the activity of the reference pentasaccharide (**47**), pointing to a prominent role of the 6-*O*-sulphate group of the H unit. PETITOU *et al.* (*70*) demonstrated that the synthetic pentasaccharide (**54**) lacking the 2-*O*-sulphate group of iduronic acid showed a similar biological activity. The protected disaccharide (**66**) bearing a p-anisyl protective group (PETITOU *et al.* (*71*)) was used for this synthesis. Surprisingly, the pentasaccharide (**55**) lacking both the 6-*O*-sulphate at H and 2-*O*-sulphate at 6 was not less active than (**53**) or (**54**). It may be concluded that neither the 2-*O*-sulphate at G nor the 6-*O*-sulphate at H is absolutely required for activation of AT III. Apparently these two sulphate groups co-operate to express maximal activity (*70*).

With regard to the role of *N*-sulphate groups at glucosamine units D, F and H RIESENFELD *et al.* (*64*) showed that removal of the *N*-sulphate at

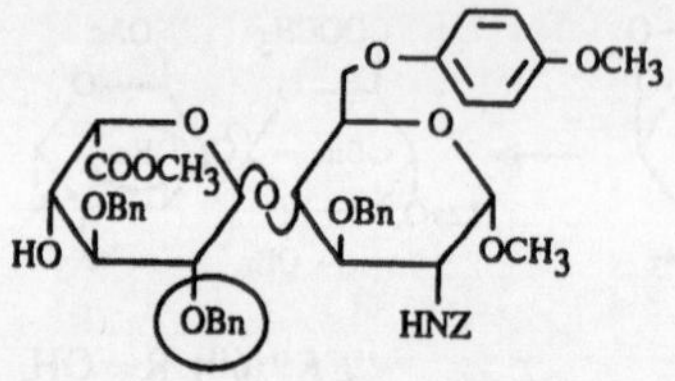

"GH" (66)

Fig. 19. Building block (66) bearing a p-anisyl ether as temporary protection, was used for the preparation of (54)

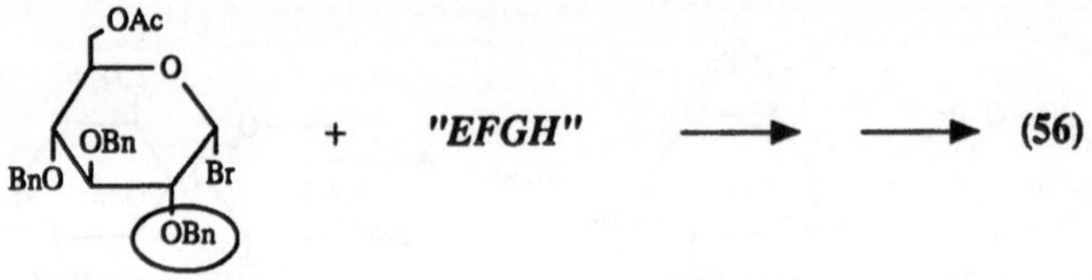

"D" (67)

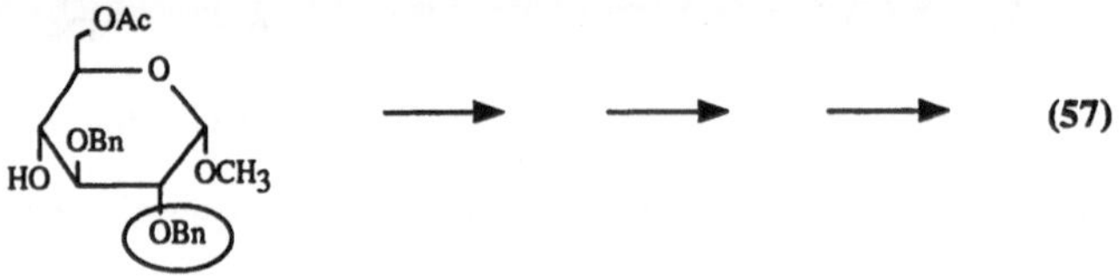

+ "EFGH" $\longrightarrow$ $\longrightarrow$ (56)

"H" (68)

$\longrightarrow$ $\longrightarrow$ $\longrightarrow$ (57)

Fig. 20. Building locks (67) and (68) have been used for the synthesis of analogues (56) and (57) containing hydroxyl instead of N-sulphate groups at unit D and H, respectively

either unit F or H leads to inactive heparin fragments, while DUCHAUSSOY et al. (61) showed (see Sect. 2.7.2), with the aid of synthetic compounds, that replacement of N-sulphate at unit D by N-acetyl leads to about 50% reduction of biological activity. Two synthetic pentasaccharides containing a hydroxyl group instead of a N-sulphate group have been synthesized (72). The activity of compound (56) containing a 2-hydroxyl group at unit D is virtually the same as that of the naturally occurring fragment with N-acetyl at D. The synthetic analogue (57) with a 2-hydroxyl group at unit H shows only 5% of the AT III mediated anti-factor Xa activity of the reference compound (47). This low activity

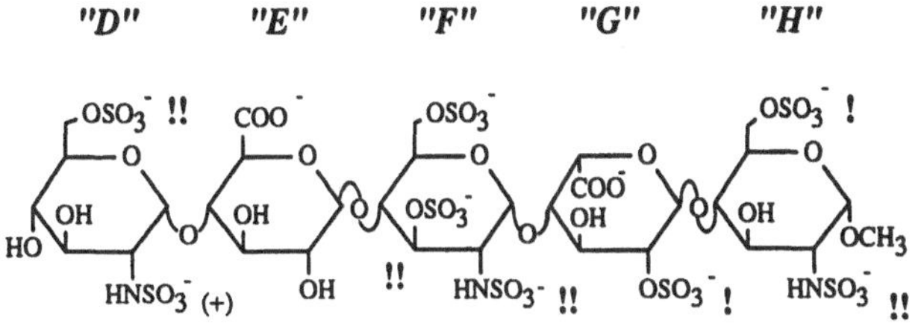

Fig. 21. The role of individual sulphate groups in the activation process of AT III. The sulphate groups indicated with exclamation marks are critical since their removal leads to more than 95% (!!) or 75% (!) loss of αXa activity; the sulphate indicated by (+) contributes weakly

corresponds to the observations by RIESENFELD *et al.* (*64*) who reported loss of activity on selective *N*-desulphation at unit H. For the synthesis of the hydroxyl analogue (**56**) a suitably protected monosaccharide (**67**) (Fig. 20) had to be incorporated containing a benzyloxy function instead of an azide group. The non-participating 2-benzyloxy function of compound (**67**) on the one hand promotes the required α-glycosidic bond formation, while on the other hand it is the appropriate function for obtaining a free hydroxyl group in the end product.

Synthesis of analogue (**57**) is similar to that of the pentasaccharides described in Sect. 2, except that for unit H building block (**68**) (Fig. 20) was used.

In conclusion, the role of the sulphate substituents of the active heparin pentasaccharide in the activation process of AT III is known precisely through studies with synthetic and biochemically modified heparin fragments. The influence of the individual sulphate groups of the pentasaccharide on the biological activity is indicated schematically in Fig. 21.

3.2. Analogues with Modifications of the Carboxylate Groups at Defined Positions

Not only the role of the sulphate groups, but also the role of the two carboxylate functions in the active heparin fragment can be studied with the help of synthetic model compounds shown in Fig. 22. From studies of AGARWAL and DANISHEFSKY (*73*) it was already known that the AT III mediated anti-Xa activity of heparin decreases upon methylation of the uronic acid carboxylate function. However, since the heterogeneous reaction products were not analyzed the drop in activity might conceivably have been caused by other modifications as for instance the cleavage of vital sulphate groups.

The first synthetic analogue lacking the carboxylate group of iduronic acid (i.e. compound (69) was described by van Boeckel et al. (74). The synthesis of compound (69) (Fig. 22), involved introduction of a 2-O-sulphated β-D-xylopyranose moiety instead of 2-O-sulphated α-L-idopyranuronate. Thus, analogue (69) had to be prepared from the protected pentasaccharide (79) in Fig. 23 following the same strategy described in previous syntheses.

Assemblage of (79) (Fig. 23) required coupling of GH disaccharide (78) with the known disaccharide (14) and monosaccharide (11), successively. A suitably protected D-xylopyranoside saccharide to be used in the synthesis of the key disaccharide building block (78) should contain a 3-O-benzyl group, a 2-O-acetyl group and a temporary blocking group at

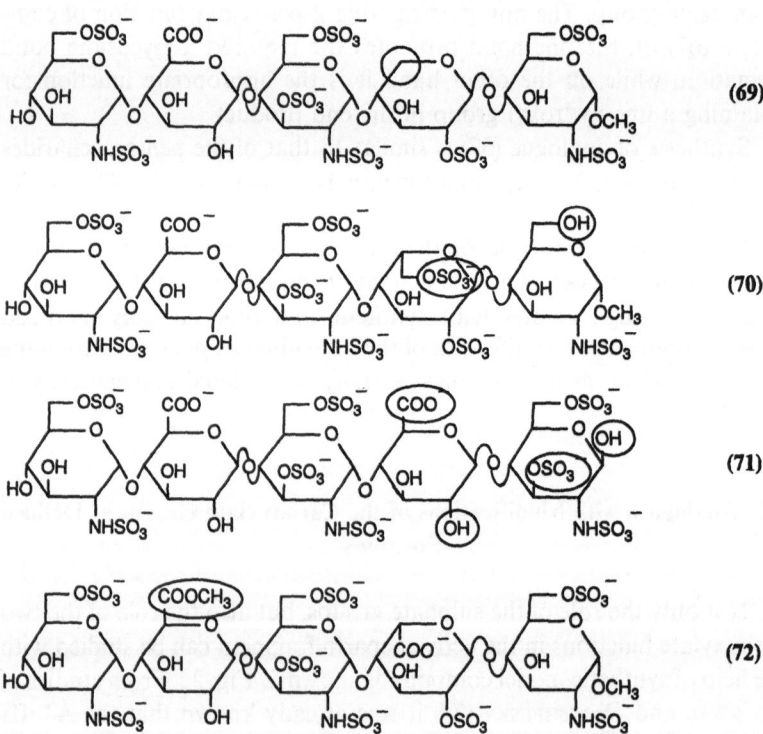

Fig. 22. Analogues with modifications at the carboxylate groups at defined positions

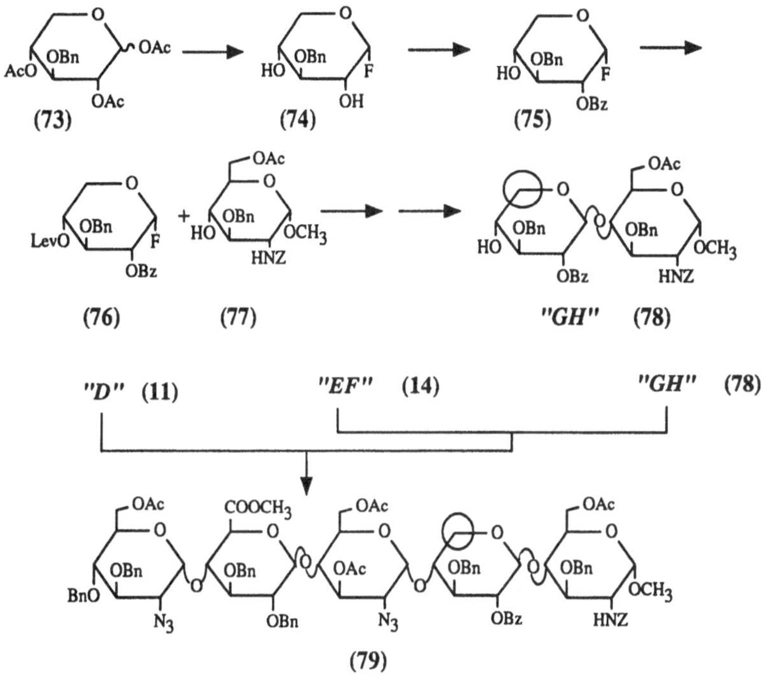

Fig. 23. Preparation of the fully protected pentasaccharide (79) for the synthesis of
analogue (69), lacking the carboxylate group of iduronic acid

the 4-O position, while in the meantime the anomeric centre should be
activated. The xylopyranoside fluoride derivative (76) fulfils these re-
quirements. The synthesis of (76) could be accomplished starting from
1,2,4-tri-O-acetyl-3-O-benzyl-D-xylopyranose (73). The latter derivative
was treated with 70% of hydrogen fluoride in pyridine and subsequently
treated with K_2CO_3 in methanol to afford (74) in 50% yield. Selective
benzoylation of the 2-OH group followed by levulinoylation gave (76) in
about 50% yield. Compound (76) was then coupled with the glucosamine
building block (77) in the presence of boron trifluoride etherate to give
after selective cleavage of the levulinoyl ester by hydrazinolysis, the
required disaccharide (78) in 42% yield.

    The pentasaccharide analogue (69) lacking the carboxylate group of
iduronic acid was found (74) to be inactive in an amidolytic assay used to
determine the anti-factor Xa activity. This sharp decrease in biological
activity may be ascribed to the loss of an essential carboxylate group

(80)

which interacts with AT III. However, the inactivity of this analogue may also be due to the fact that the xylopyranose ring displays a different conformational behaviour than iduronic acid. Thus, the α-L-idopyranuronate part of the AT III-binding fragment has been found to adopt the $^1C_4$ and $^2S_0$ conformations (see Sect. 4), whereas the 2-O-sulphated β-D-xylopyranose moiety in analogue (69) exclusively adopts the $^4C_1$ conformation.

Most likely, the carboxylate group of iduronic acid is important both for governing the conformation of iduronic acid (thereby determining the overall shape of the pentasaccharide molecule) and as a crucial charged group to interact with AT III. This hypothesis is further endorsed by the fact that pentasaccharide (70), containing 2,6-di-O-sulphated α-L-idopyranose, is biologically inactive, although it displays conformational properties similar to those of the natural product (75). Furthermore, the importance of the iduronic acid moiety is revealed by the biologically inactive analogue (71) containing D-glucuronic acid (displaying the $^4C_1$ conformation) instead of L-iduronic acid (76).

The carboxylate group of the glucuronic acid moiety (unit E) of the heparin fragment should also be involved in the AT III activation process, since pentasaccharide (72) which is selectively methylated at the glucuronic acid moiety elicits only 5% of the activity of the natural product. This pentasaccharide was obtained (77) as a side product during the preparation of (47) from the fully protected pentasaccharide (46). Apparently the methyl ester of the glucuronic acid in (46) is a stable ester function that saponifies only sluggishly as was confirmed by the isolation of intermediate (80) after saponification. It can be concluded that the two carboxylate groups of uronic acids (units E and G) in the heparin fragment are both essential for AT III activation.

3.3 A Potent Analogue with Extra 3-O-Sulphate Group at Unit H

After the role of the individual charged groups of the heparin pentasaccharide had been assessed and conformational analysis studies had been carried out, two distinct AT III binding sites could be mapped

Binding Site II

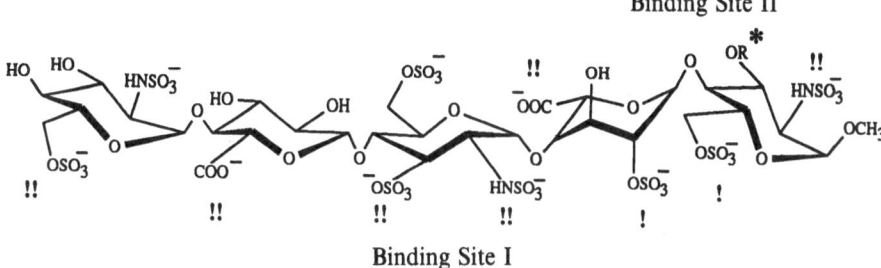

Binding Site I

Fig. 24. AT III-pentasaccharide interaction model displaying two binding sites. The interaction with binding site II could be enhanced by placing an extra 3-O-sulphate group at the reducing end (R* = SO_3^-)

(74, 78, 79) around the molecular model of the pentasaccharide as shown in Fig. 24. It can be seen that the essential charged groups, being potential AT III interaction points, are present at the south and north site of the molecule.

VAN BOECKEL et al. (78, 79) postulated that at the north site an extra O-sulphate group could be introduced to enhance the binding capacity with AT III. This extra sulphate group was introduced at the 3-O-hydroxyl group of unit H (indicated by a star in Fig. 24) to give the extra sulphated analogue (81) (Fig. 25). This analogue not only displayed a higher AT III mediated anti-factor Xa activity than the reference compound (47) (1250 U/mg vs. 700 U/mg), but also one order of magnitude stronger binding with AT III than the reference compound (80, 81). Moreover, in pharmacokinetic studies MEULEMAN et al. (59) showed that the extra sulphated analogue (81) displayed a longer biological half life compared with that of the natural compound, probably due to its higher affinity towards the AT III protein in circulation. Later PETITOU et al. (82) put forward an alternative explanation for the high activity of pentasaccharides having an extra 3-O-sulphate group at unit H. In their view anionic clustering plays an important role (70). The potent analogues possess two of these clusters (i.e. the tri-sulphated glucosamine units F and H) in distinct sub-domains, consisting of the rigid moiety DEF and the moiety GH. The iduronic acid unit G would function as the flexible bridge between the two highly anionic clusters.

The synthesis of the extra sulphated pentasaccharide (81) was achieved by following the general strategy outlined for other heparin fragments. In order to introduce a 3-O-sulphate group at the reducing glucosamine unit H a 3-O-acetyl group instead of a 3-O-benzyl group should be present in the reducing glucosamine moiety. Thus, methyl-4, 6-O-benzylidene-2-benzyloxycarbonylamino-2-deoxy-α-D-glucopyranoside was acetylated in pyridine-acetic anhydride to afford (82), which

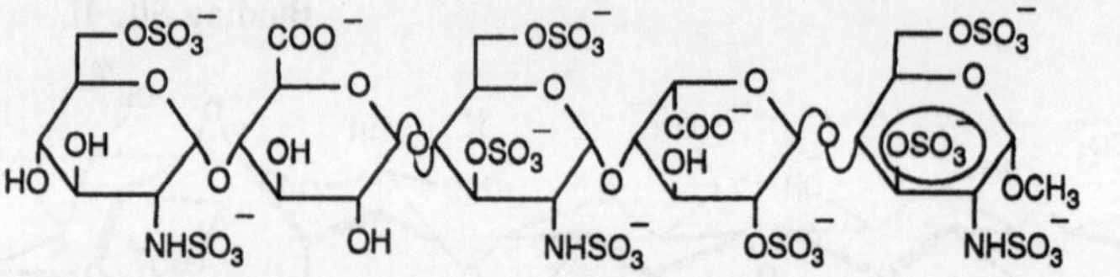

(81)

Fig. 25. The extra 3-O-sulphate group at the reducing end of (81) increases the biological activity

after treatment with aqueous acetic acid followed by selective benzoyl-ation (benzoyl chloride in pyridine at $-25°$) gave building block (83) in approximately 70% yield.

Coupling of acceptor (83) with idopyranosyl bromide (21) under conditions described for the synthesis of disaccharide (23) gave poor results, probably owing to the low reactivity of the acceptor (83) which is deactivated by the electron-withdrawing ester functions. Fortunately, coupling of the idopyranosyl fluoride (85), prepared by reaction of the acetyl derivative (84) with hydrogen fluoride-pyridine, with the glucosa-mine moiety (83) and using boron trifluoride etherate as a catalyst, afforded disaccharide (86) in 73% yield (Fig. 26).

Conversion of (86) into the desired building block (89) GH was performed by the route described for the preparation of the disaccharide blocks (14) and (24) in the synthesis of the naturally occurring fragment (see Sect. 2). Thus, disaccharide (86) was successively saponified, protec-ted at its 4'- and 6'-hydroxyl groups by way of an isopropylidene, acetylated (pyridine-acetic anhydride) and treated with aqueous acetic acid to afford intermediate (87) in 83% overall yield. Subsequently, the primary hydroxyl group of (87) was selectively protected with a dimeth-oxytrityl group. Acylation with levulinoic acid anhydride in pyridine followed by acid treatment resulted in the formation of compound (87) in 85% yield. Finally, Jones oxidation of compound (87) followed by diazomethane treatment and hydrazinolysis gave the desired building block (89) in 67% yield.

Initially the fully protected pentasaccharide was prepared (Fig. 27) by coupling of the GH block (89) with, successively, EF bromide (14) and monosaccharide bromide (11) in the presence of silver triflate as pro-moter to give α-glycosidic linkages. The fully protected pentasaccharide could be isolated in 40% overall yield. Later it was found by Basten, Jaurand et al. (83) that preparation of the fully protected pentasacchar-ide was more convenient by coupling the imidate activated DEF trisacc-haride (49), with the GH disaccharide (89), as already described for other sequences (Sect. 2.7.1). The coupling reaction was performed in dichlor-

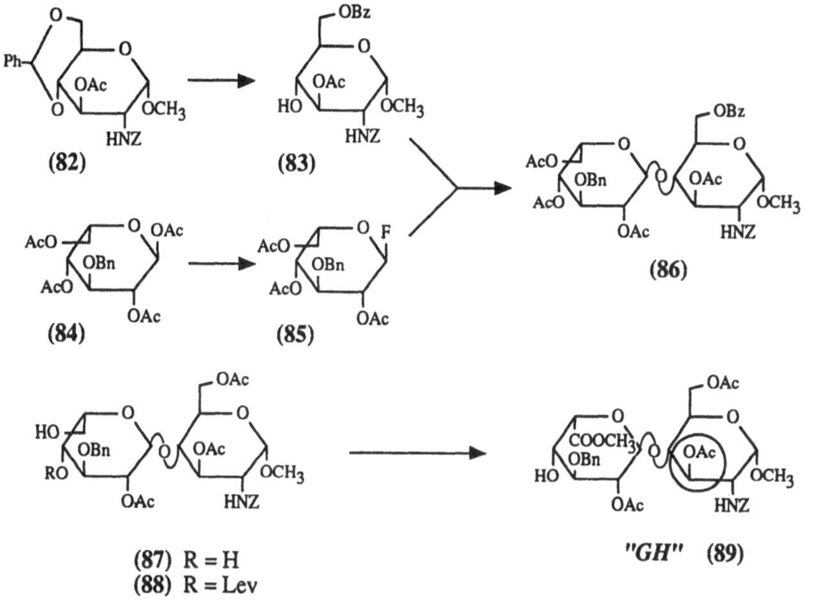

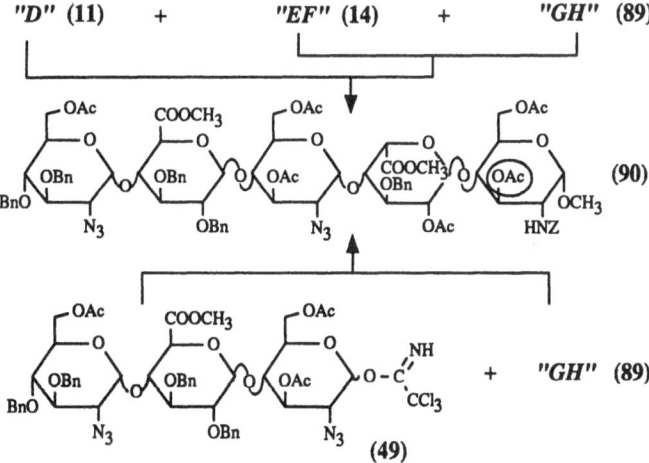

Fig. 26. Preparation of the GH building block (89) for the synthesis of the potent analogue (81)

Fig. 27. Two approaches towards the block-synthesis of the fully protected pentasaccharide (90), corresponding to the potent analogue (81) omethane using equimolar amounts of donor (49) and acceptor (89) and in the presence of trimethylsilyl triflate as catalyst (SCHMIDT (26)) to give the fully protected pentasaccharide (90) in 72% yield. Quite remarkably, the imidate coupling reaction only delivered α-coupled product, irrespective of the anomeric composition (α/β ratio) of the imidate donor (22).

Preparation of the DEF trisaccharide (49) (Fig. 28) by BASTEN et al. (84) involved a similar imidate coupling of the D monosaccharide (91) with EF building block (92) to give the corresponding trisaccharide in 75% yield. However, in this coupling reaction the undesired β-coupled product was also formed (10% yield) and had to be separated by column chromatography. The α-coupled trisaccharide obtained in this fashion was oxidized under JONES conditions and esterified with methyl iodide in DMF in the presence of potassium carbonate to give trisaccharide (93) in 58% yield. The desired DEF block (49) was then obtained in 76% yield after a three step procedure including acetolysis, selective deacetylation at the anomeric centre and subsequent reaction with trichloroacetonitrile in the presence of potassium carbonate. The fully protected pentasaccharide (90) was converted into analogue (81) by a four step procedure as described in Sect. 2.6. However, in the first step, saponification with sodium hydroxide led to approximately 40% cleavage of the pentasaccharide owing to a 4,5 elimination at the iduronic acid linkage (Fig. 29). This side reaction could be essentially suppressed by conducting the saponification with lithium hydroperoxide, generated in situ by adding lithium hydroxide to a solution containing hydrogen peroxide. After work up the saponified product (94) could be isolated in 86% yield, with

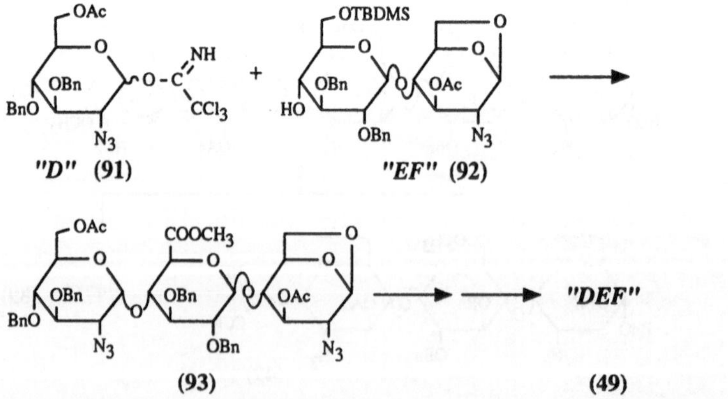

Fig. 28. Preparation of a DEF building block (49), to be used for the synthesis of various heparin pentasaccharides (see also (55))

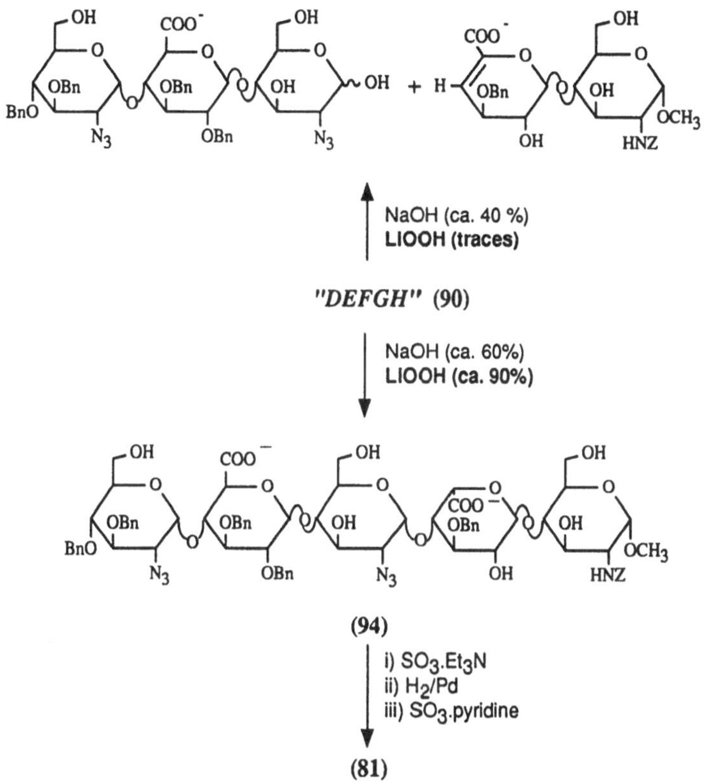

Fig. 29. Conversion of the fully protected pentasaccharide (90) into the potent analogue (81), involved a saponification step first to give intermediate (94). During the saponification cleavage of the pentasaccharide may occur between units F and G (see upper part)

elimination having been minimized. Intermediate (94) was then O-sulphated with excess of triethylamine-sulphur trioxide complex in DMF at 55 °C. After neutralization and filtration over silica gel, hydrogenolysis was performed in the presence of palladium on charcoal. Finally, selective N-sulphation could be realized with pyridine-sulphur trioxide complex in water at pH 9.5 to afford, after Sephadex chromatography (G-25), the required pentasaccharide (81) in 76% overall yield from (94).

3.4 Analogues of the Extra Sulphated Potent Analogue (81)

Because of the good biological profile of the synthetic heparin fragment with an extra 3-O-sulphate group at the reducing glucosamine unit (i.e. compound (81)), attention was turned to some of its analogues.

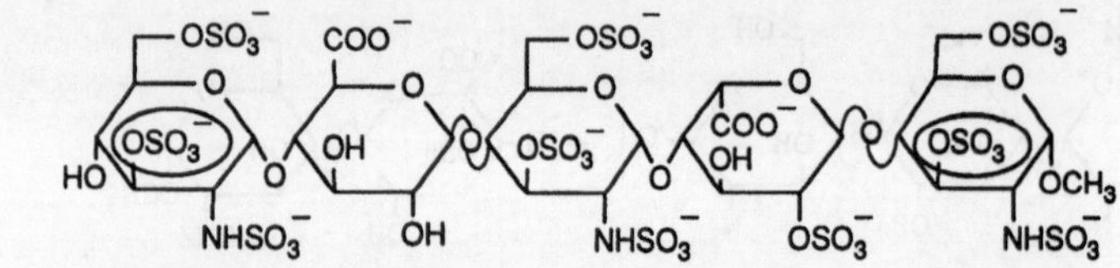

Fig. 30. Various analogues with extra 3-O-sulphate group at unit H

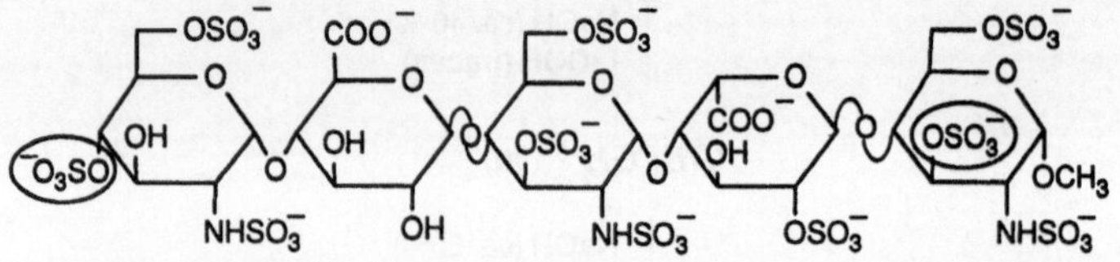

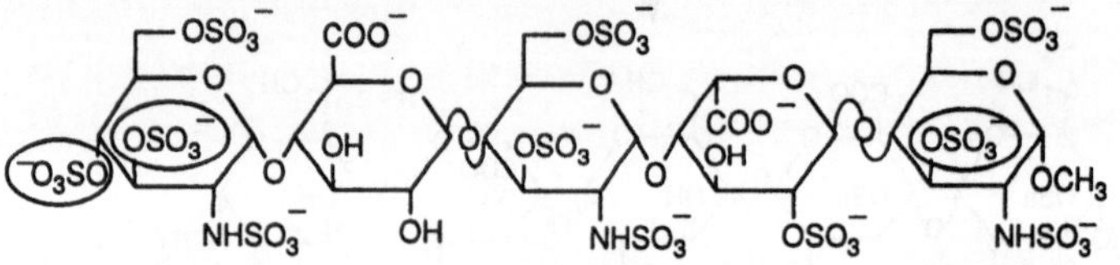

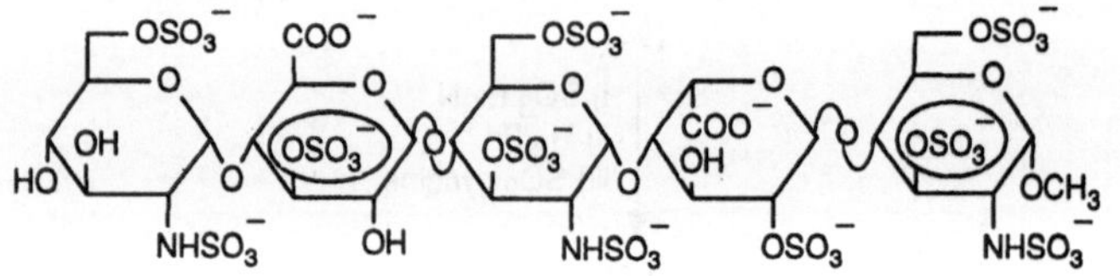

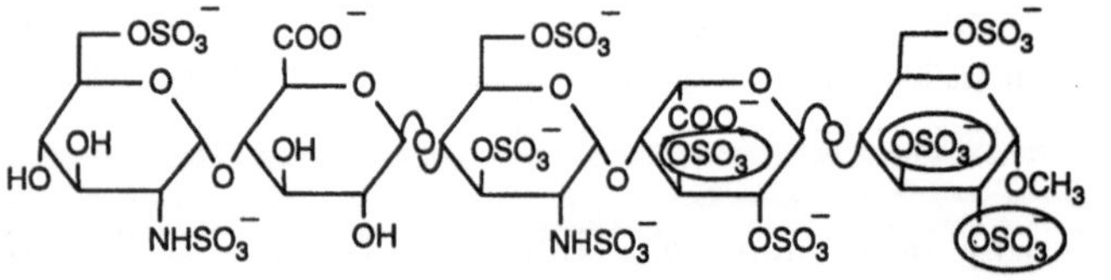

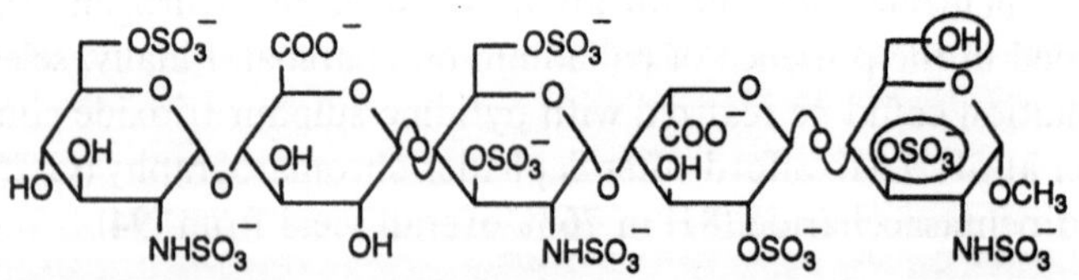

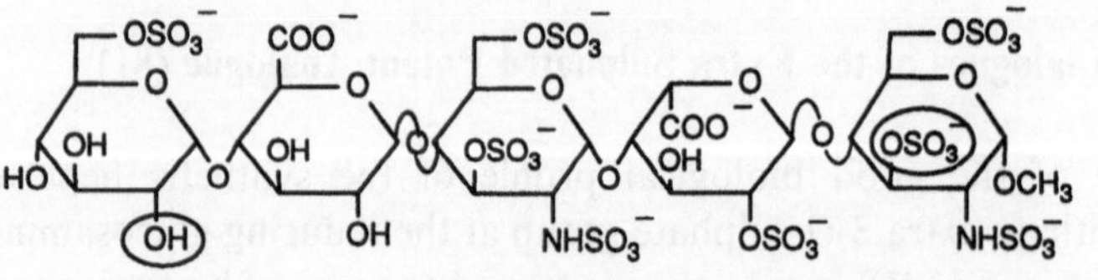

First, an answer was sought to the question whether introduction of additional sulphate groups would increase the biological activity further (*85*) unfortunately, this was not the case. Thus, analogues with one or two extra *O*-sulphate groups at unit D (i.e. compounds (**95**), (**96**) and (**97**) are slightly less active (Fig. 30). Moreover, analogue (**98**), containing an extra 3-*O*-sulphate at glucuronic acid (unit E), displays only 40% of the AT III mediated anti-factor Xa activity of the potent analogue (**81**) (*85*). When the extra 3-*O*-sulphate group is placed at iduronic acid (i.e. compound (**99**)) about 35% activity is lost with respect to the parent compound (*86*). These findings point out that activity of the heparin fragment cannot simply be increased by extra *O*-sulphation. Apparently, only the extra 3-*O*-sulphate group at glucosamine unit H enhances the interaction with a unique domain of the AT III protein.

VAN BOECKEL *et al.* (*85*) also experimented with removal of particular sulphate groups from (**81**) and its effect on the anti-factor Xa activity. The analogue (**100**) lacking the 6-*O*-sulphate group of unit H is 40% less active than the parent compound (**81**). Replacement of the *N*-sulphate at unit D by a hydroxyl group (i.e. compound (**101**)) leads to about 30% loss of anti-factor Xa activity. It is noteworthy that removal of sulphate groups from the potent analogue (**81**) has less negative impact on the AT III activation process than in the case of the reference compound (**47**).

Analogues in which N-Sulphate is Replaced by O-Sulphate

With a view to simplify the preparation of active pentasaccharides, PETITOU *et al.* (*55*) synthesized analogue (**102**) (Fig. 32). The corresponding protected H unit (**104**) was obtained in a single step from methyl α-D-glucopyranoside by treatment with benzoyl chloride in pyridine at low temperature (Fig. 31). After coupling of monosaccharide (**104**) with L-idopyranose derivative (**103**), disaccharide (**105**) could be prepared according to well-known procedures. The GH disaccharide (**105**) and the

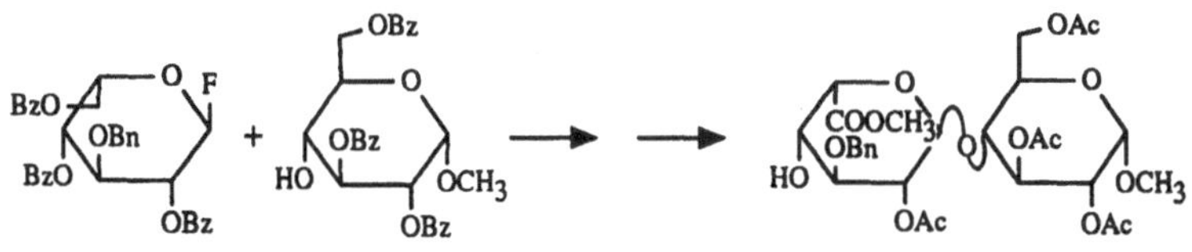

Fig. 31. Preparation of the GH disaccharide (**105**) for the synthesis of the potent analogue (**102**)

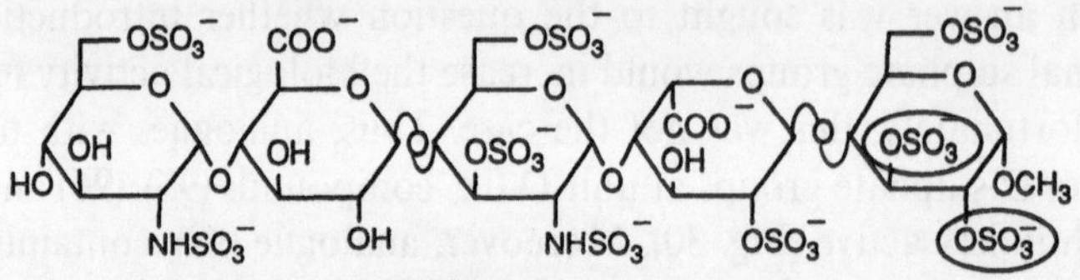

(102)

Fig. 32. In the potent analogue (102) the N-sulphate group at unit H of (81) is replaced by O-sulphate

DEF trisaccharide (49) could then be linked as described for the synthesis of (90) in Sect. 3.3. After deprotection and sulphation pentasaccharide (102) was obtained. This compound displays as high an activity as its N-sulphated counterpart (81). This result showed the way to further simplified analogues in which all N-sulphates are replaced by O-sulphates (Sect. 3.7), thereby allowing much simpler syntheses and avoiding problems arising from the introduction of N-sulphates.

3.5 Analogues Containing "Opened" Uronic Acid Moieties

Studies with synthetic heparin fragments clearly demonstrated that the carboxylate functions of the uronic acid residues E and G are strictly required for AT III binding and activation (see Sect. 3.2). The question arose if the iduronic acid moieties could be replaced by glyceric acid oxymethylene residues which expose the essential carboxylate groups in the appropriate configuration. First analogue (106) (Fig. 36) was prepared (87) in which the α-L-iduronic acid (unit G) is replaced by an R-glyceric acid 2-oxymethylene moiety (Fig. 33). In this compound the contributing 2-O-sulphate group of iduronic acid is intrinsically absent, but the extra 3-O-sulphate group at unit H may compensate for this loss (Sect. 3.3).

The crucial steps in the synthesis of analogue (106) containing an "opened" iduronic acid moiety involve synthesis and incorporation of the glyceric acid oxymethylene residue (110). The synthesis of (110) started from (R)-glyceric acid methyl ester (111). Transient protection of the primary hydroxyl group and methoxymethylation of the secondary hydroxyl group gave intermediate (112) in 70% yield (Fig. 34). Allylation of (112) followed by acetolysis (acetic anhydride, trifluoroacetic acid) of the methoxymethyl group afforded (113) in 65% yield. The latter compound was treated with hydrogen fluoride-pyridine to give the key building block (110) in 70% yield.

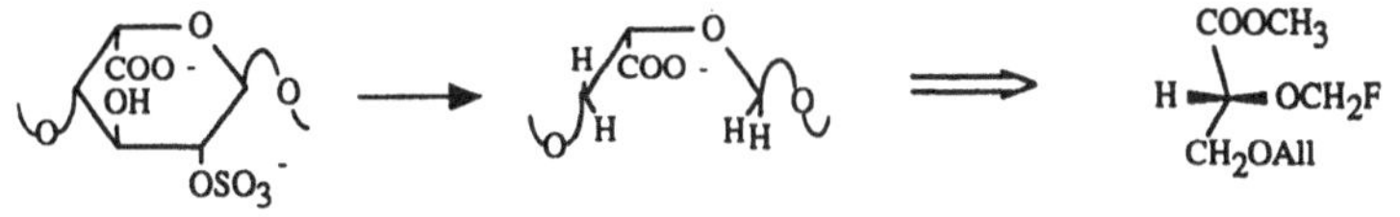

"G" (110)

Fig. 33. Replacement of L-iduronic acid by a R-glyceric acid 2-oxymethyl moiety required the preparation of building block (110)

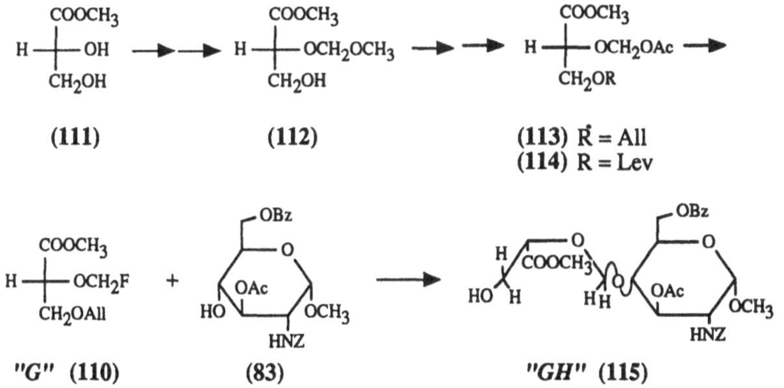

(111)   **(112)**   **(113)** R̂ = All
(114) R = Lev

"G" (110)   **(83)**   **"GH" (115)**

Fig. 34. Preparation of the GH building block (115) for the synthesis of analogues (106) and (107), containing "opened" iduronic acid moieties

Reaction of (110) with (83) in the presence of boron trifluoride etherate afforded the corresponding coupled product in 84% yield, from which the temporary allyl protecting group could be removed to give the GH dimer (115). The resultant acceptor (115) was then stereoselectively coupled with the known disaccharide donor (14) in 65% yield using mercury bromide-mercury cyanide as promoter (Fig. 35). In this particular coupling reaction a mild promoter was selected because a coupling reaction with the reactive primary hydroxyl group of (115) in the presence of active promoters, such as silver triflate, may lead to β-coupled side products (25). After removal of the temporary levulinoyl group from the resulting tetramer, monosaccharide (11) was coupled to give suitably protected pentasaccharide (116) which was then converted into the required analogue (106). Analogue (106) displayed prominent anti-factor Xa activity (about 25% of that of the reference compound), thus proving that the biological activity is retained when the complex α-L-iduronic acid is replaced by an opened "pseudo-sugar" having the carboxylate group in the appropriate configuration.

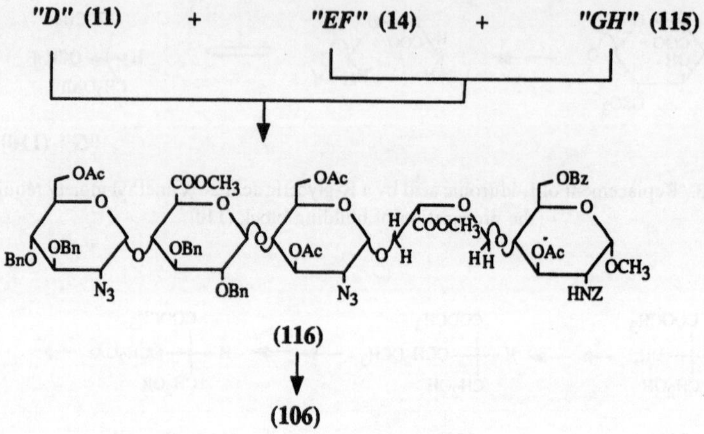

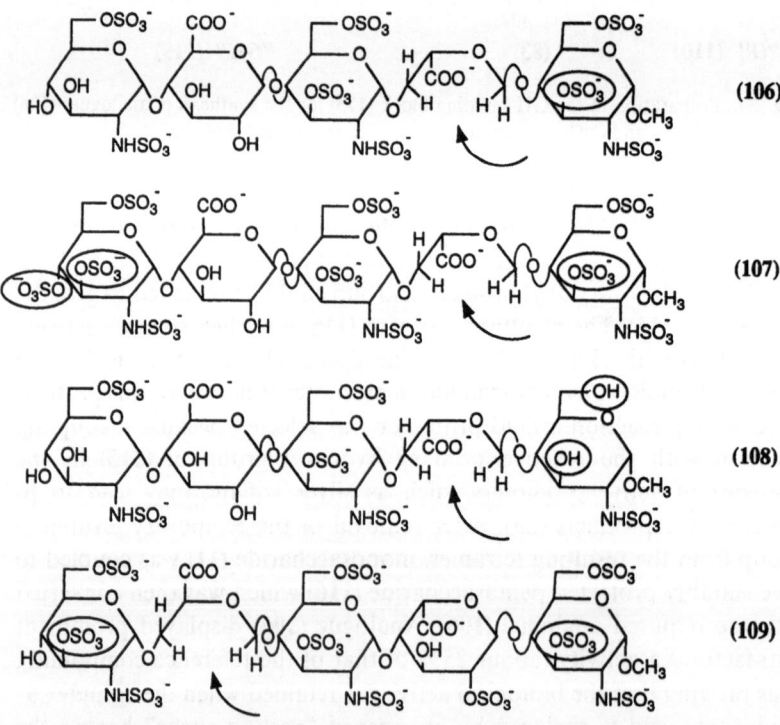

"D" (11)   +   "EF" (14)   +   "GH" (115)

(116)

(106)

Fig. 35. Preparation of analogue (106), containing "opened" iduronic acid

(106)

(107)

(108)

(109)

Fig. 36. Analogues containing "opened" uronic acid moieties

References, pp. 203–210

When compound (106) was found to be biologically active, the derivatives shown in Fig. 36. were synthesized by LUCAS et al. (88). These contain either two extra sulphate groups at the 3 and 4 positions of unit D, (i.e. compound (107)) or lack the 3 and 6-sulphate groups at unit H (i.e. compound (108)). The introduction of the extra sulphate groups in (107) does not significantly affect the anti-factor Xa activity. On the other hand compound (108) lacking the 3-O-and 6-O-sulphate groups at unit H is only slightly active, as was expected on the basis of structure-activity relationships of derivatives of the natural compound (see Sect. 3.1).

An analogue (i.e. compound (109)) containing an S-glyceric acid oxymethylene residue instead of D-glucuronic acid (unit E) was also synthesized by LUCAS et al. (88). The synthesis of this analogue started from the known disaccharide (105) followed by the stepwise elongation with suitably protected monomers (Fig. 37). Again the levulinoyl group could be successfully applied as a temporary blocking group. The protected pentasaccharide (120) was obtained in 17% overall yield from building blocks (117), (118), (119) and (105). Conversion into the required sulphated compound (109) involved previously discussed well-established routes. The analogue thus obtained was as virtually not active,

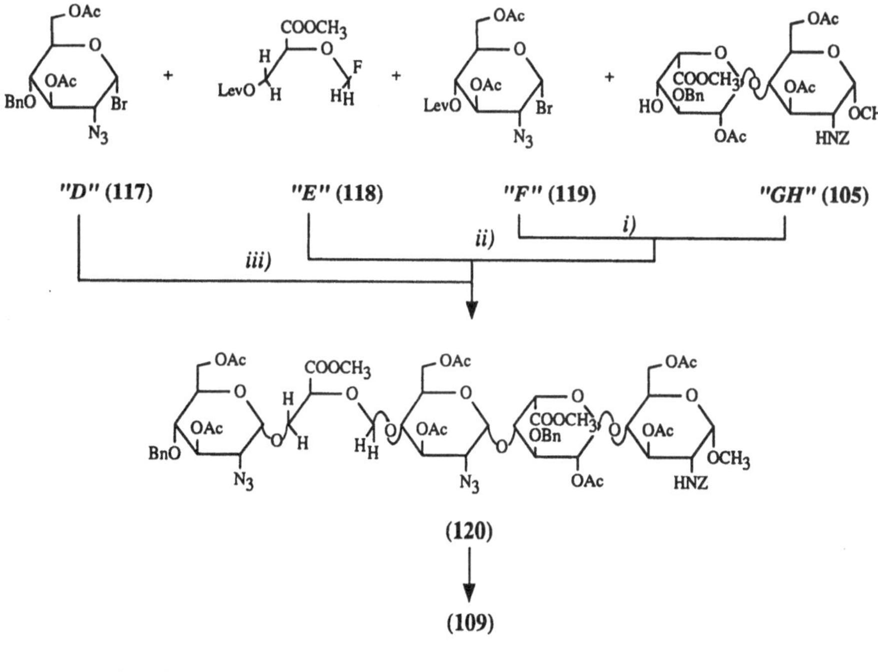

Fig. 37. Preparation of analogue (109) containing "opened" glucuronic acid

despite the fact that all essential charged groups are present in the molecule.

Apparently, the introduction of "opened" uronic acid functions in heparin like analogues is only allowed at the level of iduronic acid, known to be a flexible carbohydrate (see also Sect. 4). The glucuronic acid on the other hand is fixed in the rigid $^4C_1$ chair conformation. Most likely, the replacement of the rigid glucuronic acid by the flexible S-glyceric acid oxymethylene residue brings about a drastic change in the conformational behaviour of the heparin fragment which impedes interaction with AT III.

3.6 Analogues with Various Modifications

Compound (121) is a stereoisomer of the natural pentasaccharide fragment with the non-reducing end glucosamine (unit D) β-coupled, instead of α-coupled. This analogue was obtained (76) after deprotection and sulphation of a β-coupled, fully protected pentasaccharide which was isolated as side product in the synthesis of our reference compound (47). The β-coupled analogue displayed little bioactivity, probably caused by the presence of traces of α-coupled product (i.e. reference compound (47)).

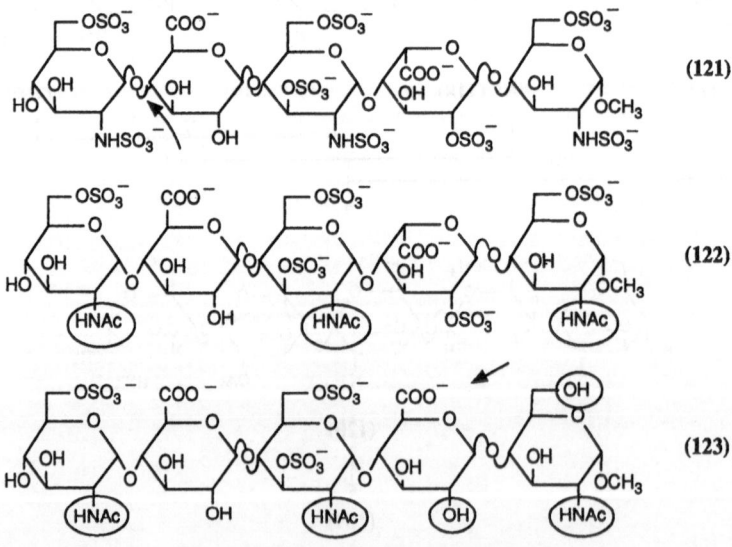

Fig. 38. Analogues with various modifications

WESSEL et al. (89) published a synthesis of the inactive tri-N-acetylated heparin pentasaccharide (122). First the known, fully protected pentasaccharide (46) was assembled using the same building blocks and coupling techniques described in Sect. 2. Instead of selective N-sulphation in the last step of the synthesis, selective N-acetylation was carried out with acetic anhydride in water to give (122). The inactive pentasaccharide (122) closely resembles the sulphated pentasaccharide (123) prepared by KRAAYEVELD and VAN BOECKEL (90). The latter derivative, however contains glucuronic acid only and is structurally related to the bacterial capsular polysaccharide isolated from E. coli K5.

Vos et al. (91) sought to discover how replacement of one of the essential sulphate groups by a more highly charged phosphate group would affect the biological activity. To this end, analogue (124) containing a 6-O-phosphate group at unit D instead of the essential 6-O-sulphate group was synthesized. The levulinoyl ester was employed for temporary protection of the 6-O-position of unit D to allow phosphorylation of this position at a later stage. The monosaccharide (125) was coupled (Fig. 39) with a suitably protected EFGH tetrasaccharide in the presence of boron trifluoride to give fully protected pentasaccharide (126) in 60% yield. Controlled hydrazinolysis of (126) gave the corresponding derivative with the free 6-hydroxyl group. Phosphorylation was then realized in 77% yield to give (127) by adding benzyl-(2-cyanoethyl) N,N-diethylaminophosphoramidite and tetrazole followed by in-situ oxidation with t-butylhydroperoxide. Saponification of the ester functions of (127) with sodium hydroxide led to concomitant cleavage of the cyanoethyl group. After O-sulphation, hydrogenolysis and selective N-sulphation, the 6-O-phosphorylated analogue (124) was purified by DEAE ion-exchange chromatography and isolated in 34% yield.

However, the phosphorylated pentasaccharide (124) displayed no significant anti-factor Xa activity. No clear explanation can be given for this difference in molecular recognition between phosphate and sulphate, although KANYO and CHRISTIANSON (92) established recently that certain proteins may discriminate between phosphate and sulphate through different networks of hydrogen bonds. From a synthetic point of view it was also attractive to prepare an analogue of the potent derivative (81) possessing a repeating unit. In compound (128), containing iduronic acid-2-O-sulphate instead of glucuronic acid, units D, F and H are alike as well as units E and G. The synthesis of compound (128) outlined in (Fig. 40) followed well-established procedures (84). In principle, the pentasaccharide (128) may be synthesized from two key building blocks designed to deliver trisulphated glucosamine moieties (corresponding to D, F and H) and monosulphated iduronic acid moieties (corresponding

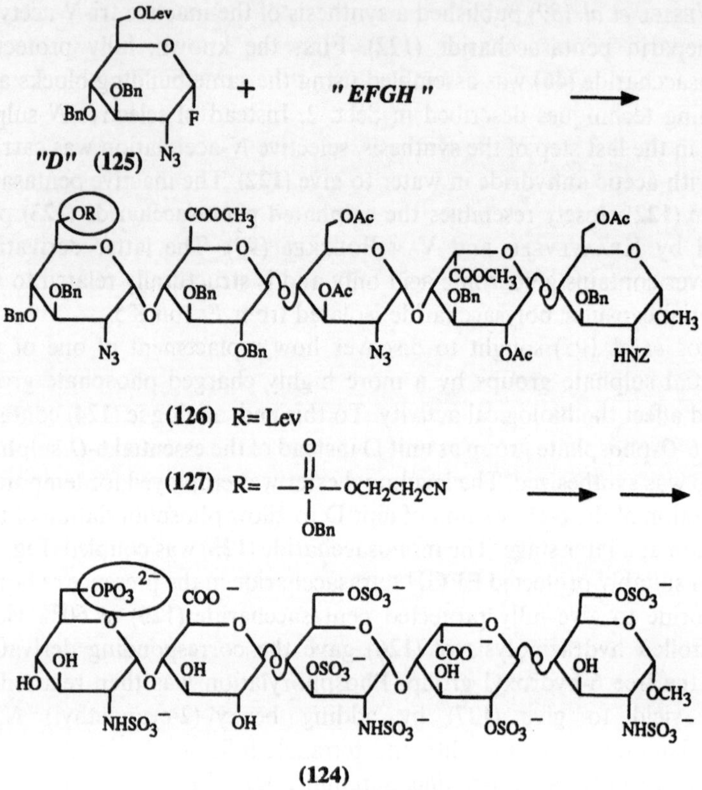

Fig. 39. Preparation of the inactive analogue (**124**), containing a 6-*O*-phosphate group
instead of 6-*O*-sulphate group at unit D

to E and G) respectively. The analogue (**128**) displays about 10% of the
anti-factor Xa activity of the potent derivative (**81**). In this respect it is
interesting to note that recently a naturally occurring heparan sulphate
of glomerular basement membrane was characterized by EDGE and SPIRO
(*93*) who found that a substantial portion of the outer region of the
polymer contains a similar repeating unit as analogue (**128**) namely
[IdoA(2-SO$_4$)α1 → 4 GlcNSO$_3$(3-SO$_4$)α1 → 4].

Since analogue (**128**) containing two iduronic acid moieties still
elicited anti-factor Xa activity and since we found previously that α-L-
iduronic acid in the active heparin fragment may be replaced by a R-
glyceric acid oxymethylene residue (see Sect. 3.5) we decided to synthesize the highly simplified pentamer (**130**). Unfortunately, this compound

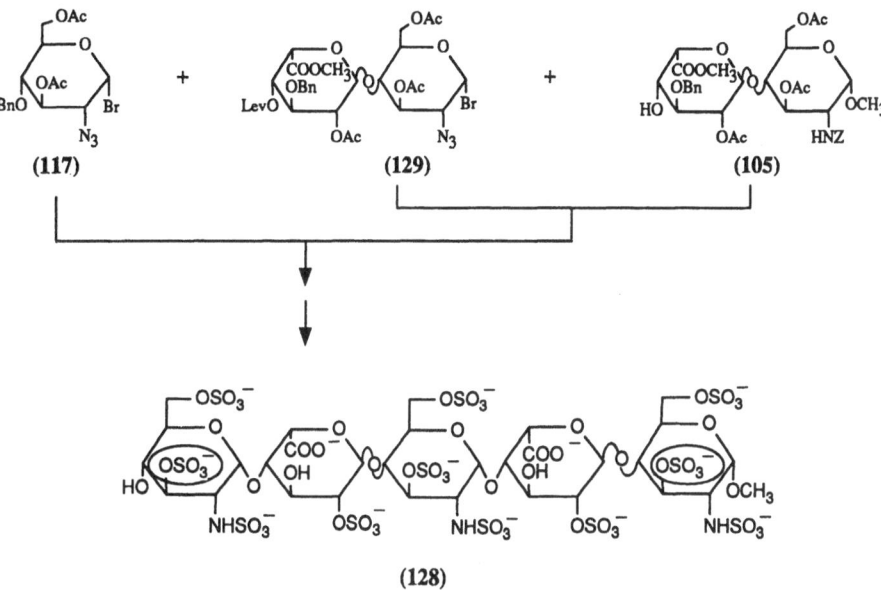

Fig. 40. Synthesis of analogue (128), displaying a repeating sequence D = F = H; E = G

is biologically inactive. In the next section we will see that this cannot be ascribed to the presence of sulphated glucose residues instead of glucosamine residues.

The synthesis of pentamer (130), (Fig. 41) involved the preparation of key intermediate (132) which was used in two consecutive steps (94). Thus, selective coupling of the primary alcohol function of R-glyceric acid methyl ester with bromide (131) under the *in situ* anomerization conditions described by LEMIEUX *et al.* (95) afforded the corresponding dimer in 58% yield. Next, the free secondary hydroxyl function of the dimer was functionalized with a pentenyl oxymethyl (POM) ether by treatment with POM-CL in the presence of a tertiary base to give building block (144) in 82% yield. The 4-pentenyl (Pent) group present in key intermediate (132) was used here as the active function on the pseudo-anomeric centre (the oxymethylene group) instead of the fluoride used in the synthesis of the "opened" analogue (106). The reagent of choice in the two coupling reactions with donor (132) was iodonium triflate, leading to approximately 65% yield of the desired coupling products. The protected pentamer (133) was then deprotected and sulphated to deliver compound (130) in 55% yield.

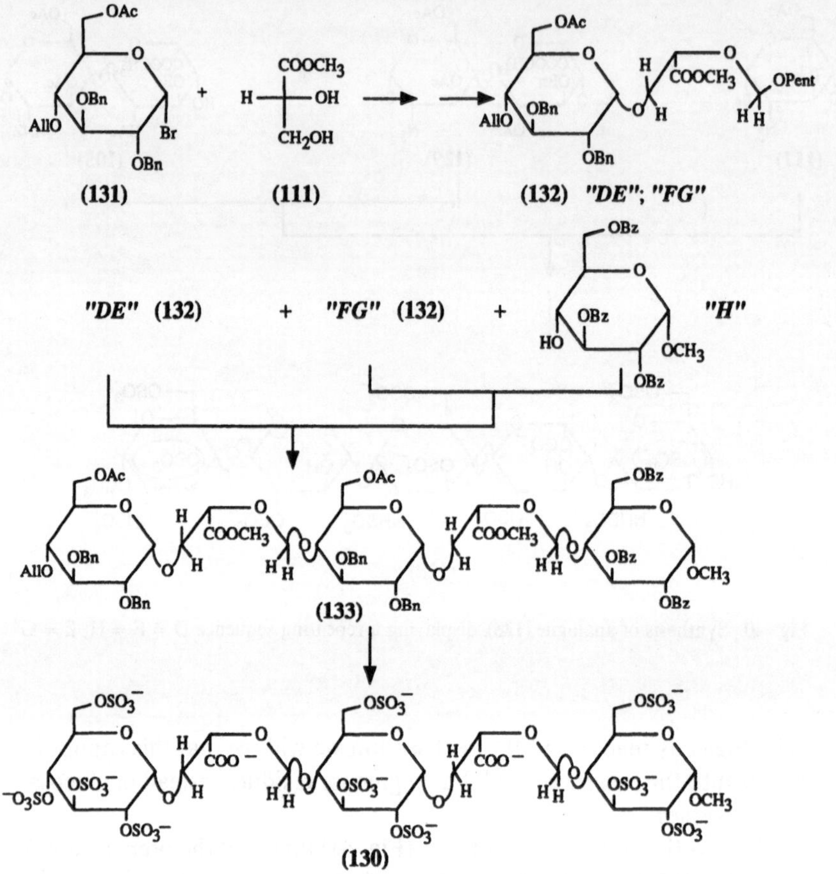

Fig. 41. Preparation of the highly simplified heparin analogue (130), containing two "opened" iduronic acid moieties

3.7 Alkylated Analogues of Heparin Pentasaccharides

In Sect. 3.1–3.5 many analogues of the naturally occurring pentasaccharide have been described. We reasoned (96, 110–112) that a completely new series of active analogues might be prepared in which all, or part, of the free hydroxyl groups of the pentasaccharide would be alkylated. First, two partially methylated derivatives of the potent pentasaccharides (102) and (81), were synthesized, i.e. compounds (134) and (135), respectively, and found to be almost as active as the non-methylated pentasaccharides (Fig. 42).

Synthesis (110) of the pentasaccharide (134) and of other derivatives having a 3-O-methyl group in the iduronic acid portion necessitated

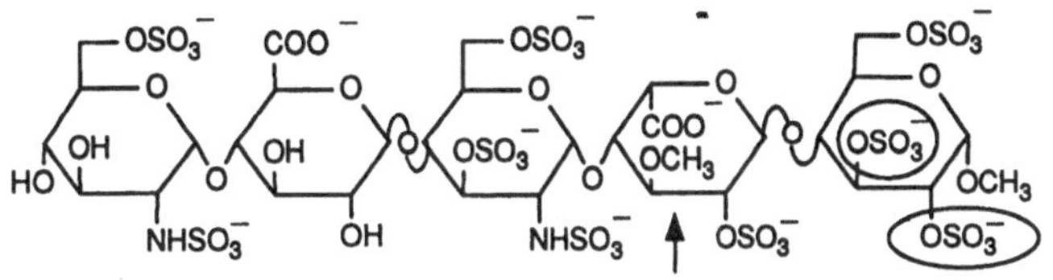

Fig. 42. Partially methylated heparin pentasaccharides

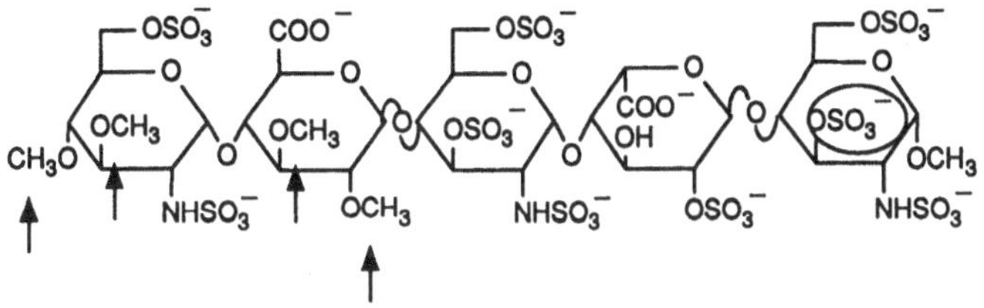

preparation of the idopyranose fluoride or thio derivative (138) or (139). These compounds can be prepared in analogy with the synthetic route to the 3-O-benzylated idopyranose derivative (85), either by starting from 1,2-5,6 di-O-isopropylidene glucofuranose or, in a shorter synthesis, using the known L-idofuranose derivative (19) which could be successively hydrogenolyzed, methylated and then converted into the required building blocks (138) and (139) (Fig. 43). Compound (138) was then coupled with the reducing end building block (104) in the presence of boron trifluoride to give the corresponding disaccharide (140) in 68% yield. In an eight-step synthesis, identical with the preparation of the benzylated analogue (105) (Sect. 3.4), disaccharide (140) was converted into the suitably protected GH block (141) in 21% overall yield (Fig. 44). Coupling of the GH-disaccharide (141) with the known DEF-trisaccharide (49) gave a fully protected pentasaccharide, which after deprotection and sulphation gave the iduronic acid methylated pentasaccharide (134), which turned out to be as active as the potent analogues described in Sects. 3.3 and 3.4.

We also decided to synthesize (111) pentasaccharide (135) in which the four hydroxyl groups of units D and E have been methylated (Fig. 42). For this synthesis the methylated DEF-trisaccharide building block (147) had to be synthesized (Fig. 45). First methylation of intermediate (143) gave disaccharide (144) in quantitative yield. Following our usual synthetic procedures, which do not call for special comment, the suitably protected EF disaccharide (146) could be obtained. For the preparation of the D monomer (145) the anhydro-glucose derivative (142) was methylated and then converted into the trichloro-acetimidate derivative.

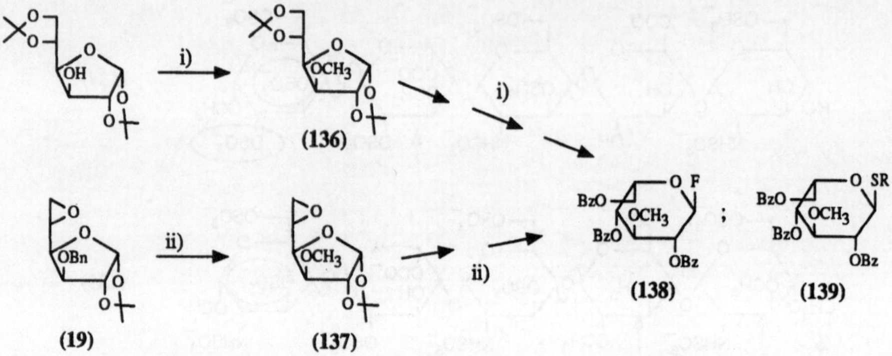

Fig. 43. Preparation of 3-O-methylated iduronic acid building blocks

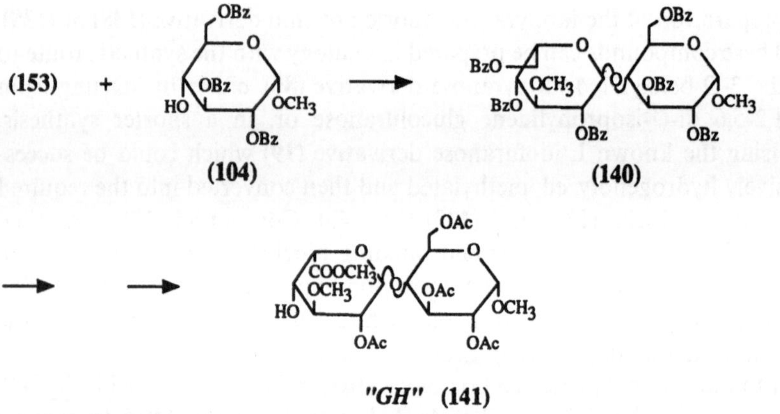

"GH" **(141)**

Fig. 44. Preparation of the GH building block **(141)** for the synthesis of O-methylated analogues **(134)**, **(148)**, **(149)**, **(150)**, **(151)**

Coupling of donor **(145)** with acceptor **(146)** in the presence of trimethyl-silyl triflate furnished an α-coupled trisaccharide, which was then sub-jected to JONES oxidation and treatment with diazomethane to give trisaccharide **(147)** in 30% overall yield from **(146)**. Having trisaccharide **(147)** at our disposal, the preparation of the methylated pentasaccharide **(135)** followed a similar synthetic route as that of the potent analogue **(81)** described in (Sect. 3.3). The methylated analogue **(135)** displayed the same favourable biological activity as the potent derivatives **(81)** and **(102)**.

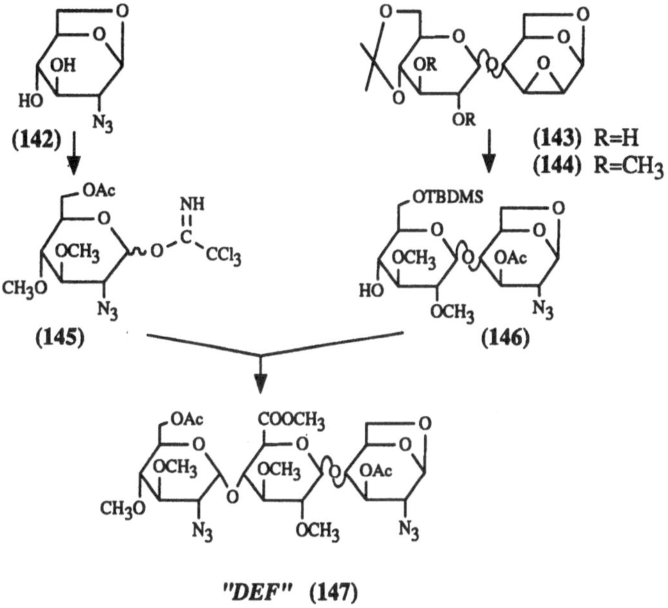

Fig. 45. Preparation of the methylated DEF building block (147) for the preparation of the partially methylated derivative (135)

Since the biological data of the methylated pentasaccharides (134) and (135) indicated that methylation of free hydroxyl groups of heparin fragments does not affect the AT III activation process and since the N-sulphate group of unit H may be substituted by an O-sulphate group (see Sect. 3.4), we were anxious to synthesize a "non-glycosaminoglycan" analogue containing solely O-sulphate esters and O-methyl ethers (i.e. pentasaccharide (148)) (Fig. 46). Fortunately, it appeared that these modifications are not at the cost of the biological activity. Eventually, after profilation of fully O-alkylated and O-sulphated pentasaccharides, some of the pharmacological parameters turned out (97, 112) to be significantly improved with respect to the potent analogue (81). More-over, the synthesis of this class of analogues is much easier than that of the naturally occurring compounds for several reasons. i) No amino-sugars have to be introduced which require elaborate synthetic routes for the preparation of azide containing building blocks; ii) the synthetic strategy is not restricted to the use of benzyl-ethers and acyl esters at defined positions, because no free hydroxyl groups are present; con-sequently, both acyl and benzyl protective groups can be selected for protection of the hydroxyl groups to be sulphated iii) fully protected

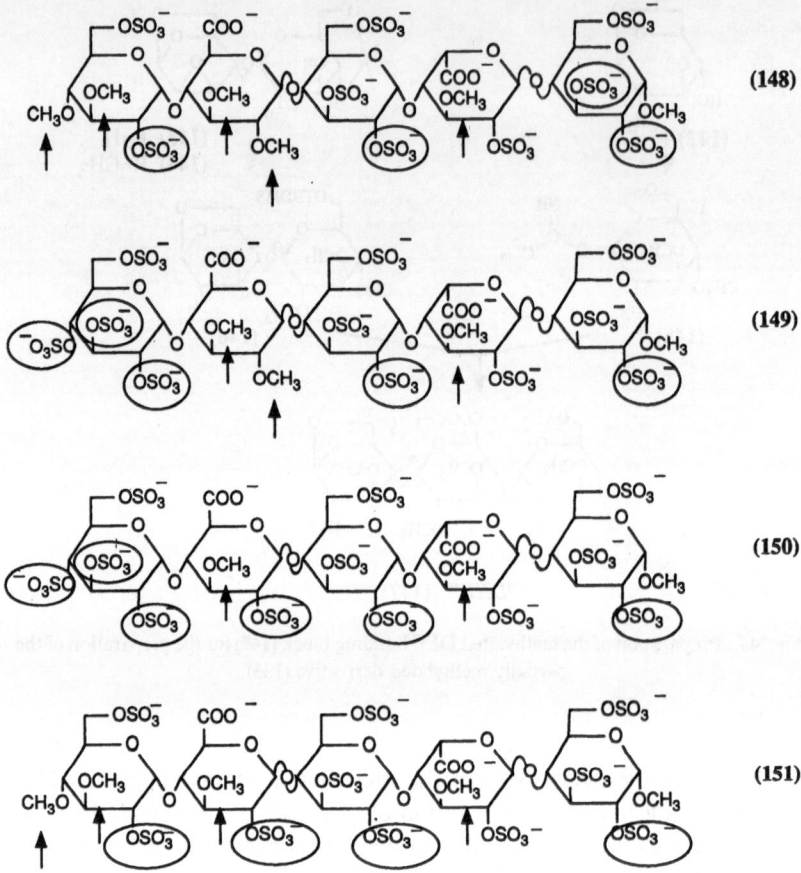

Fig. 46. Potent analogues containing only O-methyl and O-sulphate groups

pentasaccharides can be prepared in which all hydroxyl and carboxylate groups contain benzyl functions; in this way one has not only the advantage of one deprotection step but also abolishes the risk of β-elimination at the level of the uronic acids; iv) at the end of the synthesis selective N-sulphation is not necessary.

For the rapid synthesis (*112*) of compound (**148**), however, it was most convenient to use the building blocks which were already in stock. Obviously, more economical routes can easily be devised. First, the methylated disaccharide (**144**) was subjected to treatment with tetrabutylammonium hydroxide at elevated temperature to open its epoxide (Fig. 47). The crude product thus obtained was benzylated, treated with aqueous acetic acid to cleave the isopropylidine group, and then selectively protected with a t-butyldimethysilyl group to afford the EF-disaccharide (155) in 72% overall yield. Disaccharide (155) was then coupled with the easily accessible donor (153) in the presence of mercury cyanide as promoter to provide trisaccharide (156) in 62% yield.

JONES oxidation of (156), followed by methyl-esterification gave trisaccharide (157) in 56% yield. Treatment of (157) with a mixture of acetic anhydride and trifluoroacetic acid for 16 hours at room temperature not only led to acetolysis of the 1,6-anhydro ring but also to selective acetolysis of the 3-O-benzyl group at unit F (3-O-benzyl derivative (158): 27% yield; 3-O-acetyl derivative (159): 45% yield). After selective deacetylation of (159) at the anomeric centre with piperidine in THF, bromide (160) could be obtained in 81% overall yield when the anomeric hydroxyl group was allowed to react with oxalyl bromide in a

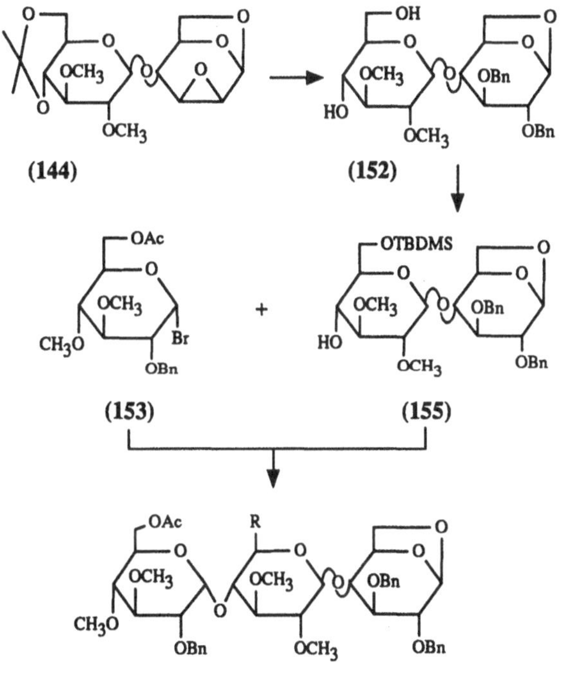

"DEF" (156) R = CH₂OTBDMS
(157) R = COOCH₃

Fig. 47. Preparation of the methylated DEF building block (157) for the synthesis of analogue (148)

mixture of chloroform and DMF. Fully protected pentasaccharide (162) was obtained in 62% yield by reaction of donor (160) with acceptor (161) in the presence of mercury cyanide-mercury bromide. The fully protected pentasaccharide was successively saponified, hydrogenolyzed and sulphated to give the required pentasaccharide (162) in 60% overall yield. As already mentioned above the O-sulphated/O-methylated analogue (148) displayed a somewhat higher AT III mediated anti-factor Xa activity and a stronger binding with AT III than the potent compound (81).

As in the series mentioned in Sect. 3.4, several extra sulphated methylated analogues were also synthesized (i.e. compounds (149), (150) and (151; Fig. 46). In the meantime, we prepared (83) analogues with an extra 2-O-sulphate group at the glucuronic acid unit E (i.e. compounds (150) and (151)). The presence of the extra sulphate groups at units D

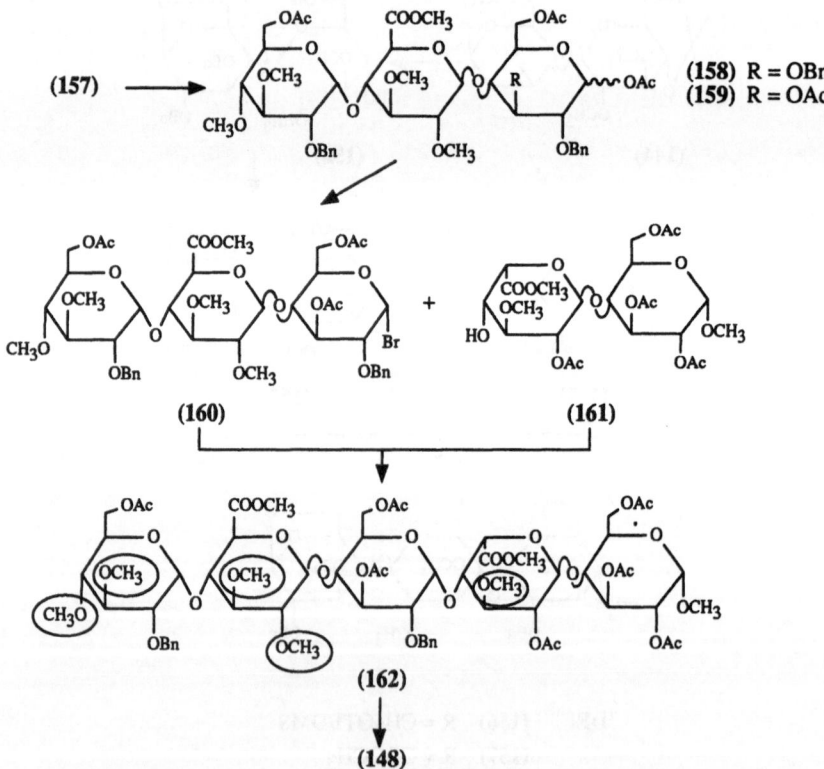

Fig. 48. Synthesis of the O-sulphated and O-methylated analogue (148), which is highly biologically active

and/or E led only to a slight reduction of the AT III mediated anti-factor Xa activity.

In addition we also synthesized (96) pentasaccharides substituted with longer aliphatic chains. Three interesting examples are illustrated in Fig. 49. Compound (163) contains two linear tetradecyl (C14) chains at unit D. This modification, however, is accompanied by a nearly complete loss of anti-factor Xa activity. On the other hand when two hexyl chains were placed at the same positions (i.e. compound (164)) the activity was restored to about 80% of the activity of the methylated pentasaccharide (148). Apparently, when the alkyl substituents at D become too long they hamper the interaction with AT III. For synthesis of the compounds with C6 and C14 alkyl chains at unit D we used as shown in Fig. 50, the same precursor as for the methylated monosaccharide (153), used for the preparation of pentasaccharide (148). Thus, monosaccharide (166) was treated with sodium in benzyl alcohol at elevated temperature to give the benzyloxy adduct which upon treatment with hydrogen chloride in aqueous dioxan was partially deblocked to give (167) in 46% yield. Then, the required building blocks (153) and (168) could be obtained in about 80% yield by alkylation of the hydroxyl groups followed by acetolysis of the anhydro ring and subsequent bromination of the anomeric centre with titanium tetrabromide.

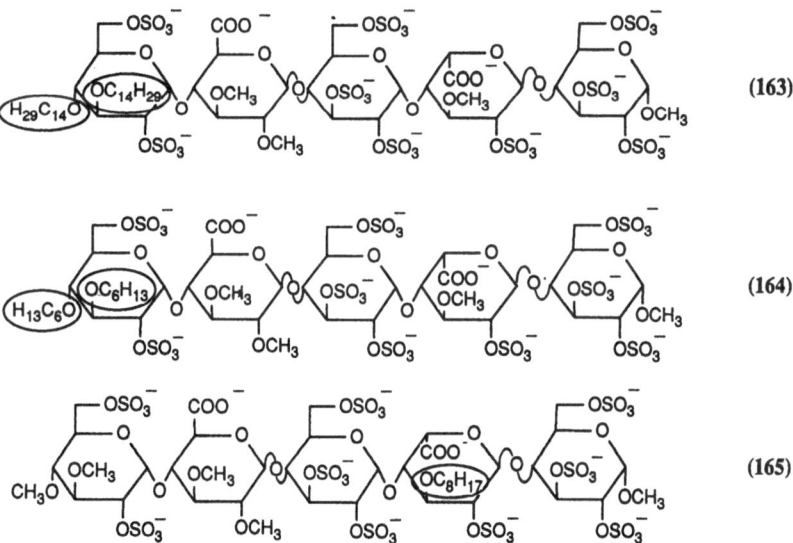

Fig. 49. Analogues which contain apart from O-methyl groups also other O-alkyl groups

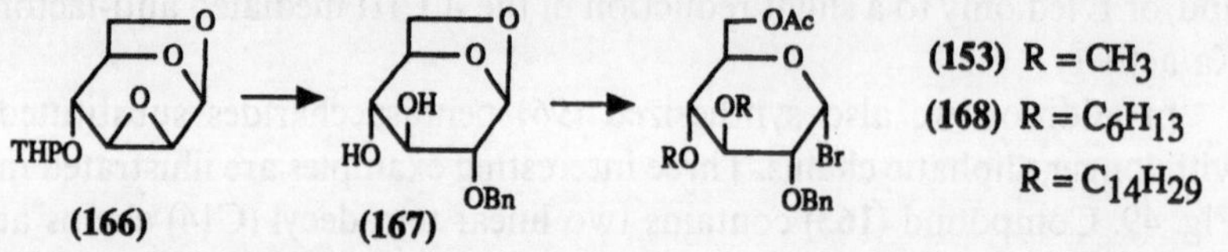

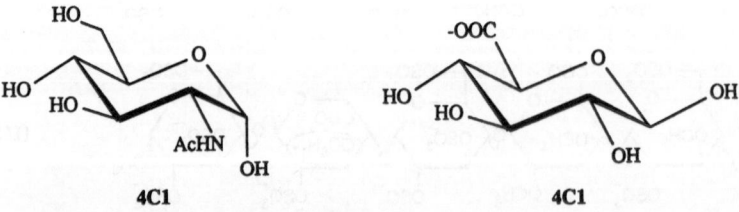

(153) R = CH₃

(168) R = C₆H₁₃

R = C₁₄H₂₉

(166)   **(167)**

Fig. 50. Preparation of 3,4-O-dialkylated building blocks corresponding to the D unit

4. Conformational Properties

The conformation of any oligo- or polysaccharide depends on the conformation adopted by the individual sugar moieties and the conformational relationships between adjacent residues. Conformational analysis of heparin fragments is complicated by the presence of the charged sulphates and carboxylates.

In heparin and heparin fragments the glucosamine and glucuronic acid residues have invariably been found to adopt the $^4C_1$ conformation in which all non-hydrogen substituents are in the energetically favourable equatorial orientation except for the axially oriented anomeric oxygen in α-D-glucosamine (Fig. 51). The occurrence of $^4C_1$ conformers is in accordance with observed interproton coupling constants of these residues (98).

The situation with respect to the iduronic acid residues is far more complex. First experimental evidence based on X-ray diffraction studies suggested that the L-iduronate residues in heparin are in the $^1C_4$ conformation and the observed ^{1}H-NMR interproton coupling constants seemed to correspond with this (98). However, recent studies on synthetic and some natural heparin fragments revealed coupling constants not in agreement with the initial supposition. Thus, in pentasaccharides (1) and (2) values of 7.54 Hz and 3.56 Hz were observed for the

Fig. 51. The most stable conformers of α-D-glucosamine and β-D-glucuronic acid residues vicinal coupling constants $J_{2,3}$ and $J_{3,4}$, respectively compared with the smaller and similar values (around 2.60 Hz) to be expected for a $^1C_4$ conformer (99). Theoretical calculations showing that a third conformer, the skew-boat conformer $^2S_0$, is equi-energetic with the two chair $^4C_1$ and $^1C_4$ conformer (100) provided an explanation for this observation; thus the idopyranuronic acid residues are in equilibrium between the $^2S_0$, $^4C_1$ and $^1C_4$ conformers (Fig. 52). As in the $^2S_0$ conformer H-2 and H-5 of the iduronate rings are close to each other (2.3 Å) the presence of this conformer could be confirmed by NMR spectroscopy through observation of a strong NOE between these protons when iduronic acid is in some environments (101, 104) as explained below.

Analysis of the coupling constants observed for the iduronate ring in various oligosaccharides revealed that they could be rationalized in terms of an equilibrium between the three conformers, the contribution of each one depending on the nature of the ring substituents, the adjacent residues, and the presence of sulphate groups (101–104). The same behaviour was found for the idose ring (105).

The influence of the sulphate at the 2 position of iduronic acid is particularly pronounced when this residue is present at the non-reducing end of an oligosaccharide (compare data on (169), (170), (171) and (172) in Fig. 53. In this case, the monomer methyl-α-L-iduronate 2-sulphate is predominantly in the $^1C_4$ conformation. However, when it is incorporated between two N-sulphate-6-sulphate glucosamine residues the $^2S_0$

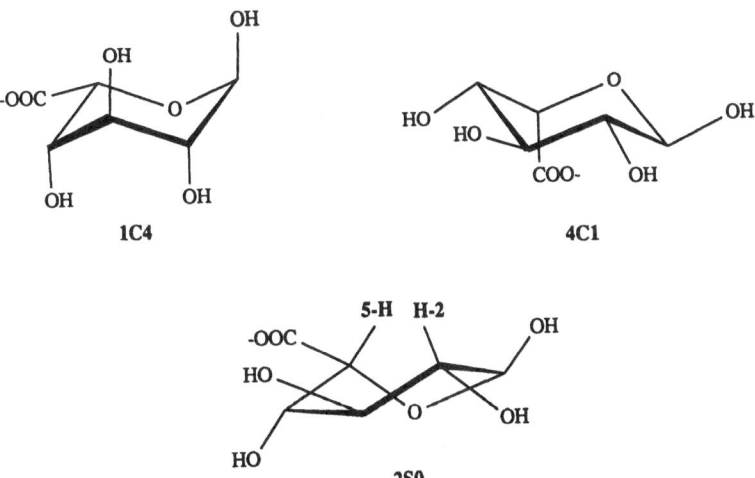

Fig. 52. The three energetically favoured conformers of L-iduronic acid residues in heparin. Note the short distance between H-2 and H-5 in the $^2S_0$ conformer

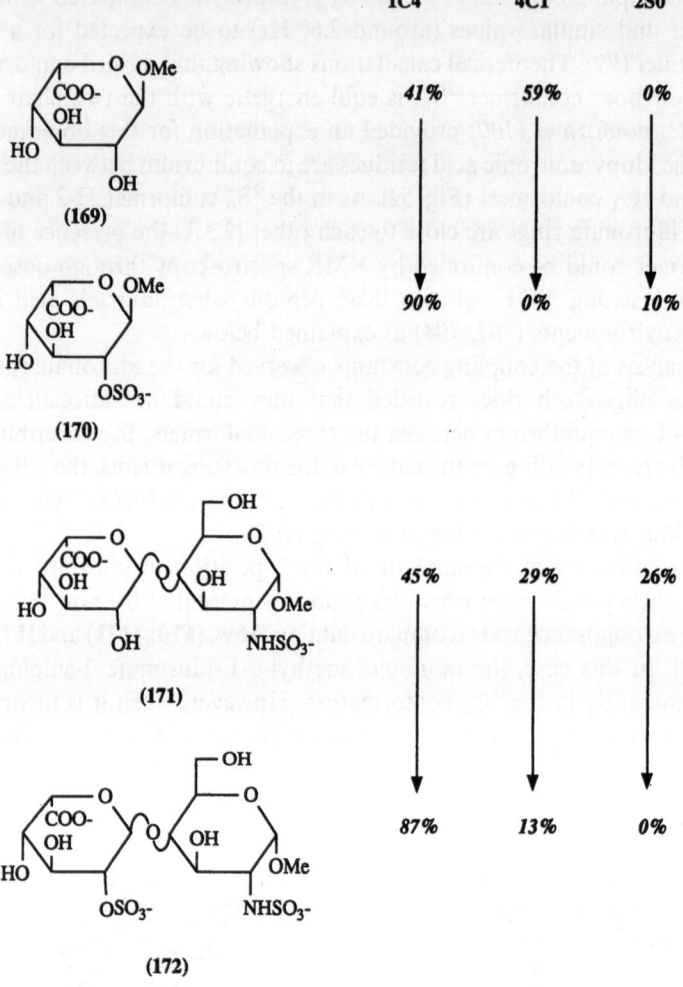

Fig. 53. Conformation of sulphated and non sulphated iduronic acid residues in different environments. The figures indicate the percentage of each contributor

conformer becomes a strong contributor to the equilibrium (30 to 50%) as is shown clearly in the hexasaccharide (173) which contains one terminal and two internal 2-O-sulphated iduronate units (Fig. 54). Introduction of a sulphate group at the 3-position of one glucosamine (in the pentasaccharide sequence responsible for the binding to AT III) leads to predominance of the $^2S_0$ conformer (99) while addition of a second sulphate group at the 3-position of the other glucosamine completely shifts the equilibrium toward the $^2S_0$ conformation (78) (Fig. 55).

References, pp. 203–210

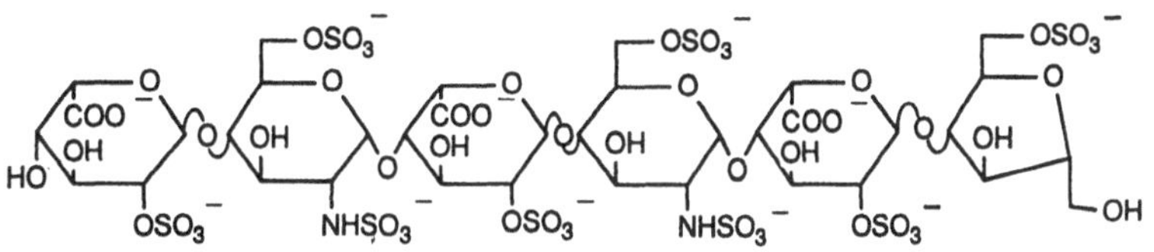

(173)

Fig. 54. Conformation of sulphated iduronic acid residues in a heparin hexasaccharide. The figures indicate the contribution of the $^1C_4$, $^4C_1$ and $^2S_0$ conformers respectively

It appears that the nature of remote substituents can also influence the conformational equilibrium of the iduronic acid portion. Thus in the N-acetylated and N-sulphated pentasaccharides corresponding to the binding sequence to AT III, the $^2S_0$ conformer is more favoured in the N-acetylated than in the N-sulphated counterpart (*61*) (Fig. 55). The actual meaning of these structure-associated conformational changes of L-iduronic acid residues is not obvious. Whether they influence the binding of heparin oligosaccharides to AT III is not proved. It is clear from biological and binding data that high affinity for AT III is associated with strong predominance of the $^2S_0$ conformer. However, the lower binding affinity of the N-acetylated pentasaccharide sequence compared to the N-sulphated one (*61*), in spite of a greater contribution of the $^2S_0$ conformer, would indicate that the initial conformation of the L-iduronate ring (before binding to the protein) is of little importance.

Another still pending question is the conformation of the iduronate ring bound to the protein. Although, as mentioned above, the strength of binding appears in most cases to be in direct relationship with the contribution of the $^2S_0$ conformer in the conformational equilibrium of the free oligosaccharide, it has been suggested (*101*), based on NMR studies at various ionic strengths, that the bound form could be only $^1C_4$. However, definitive experimental evidence is needed.

The influence of sulphate groups on the conformation of an oligosaccharide is not fully understood. It is clear that electrostatic and nonbonding interactions will effect the conformational behaviour. Furthermore, one may expect cations to influence the conformation of sulphated oligosaccharides. However, it was found in one study, using model compounds, that neither the sulphate nor the counter ion had an influence on the conformational equilibrium of the hydroxymethyl group (*106*). However sulphation at the 6-position of glucosamine tends to

	1C4	2SO	4C1

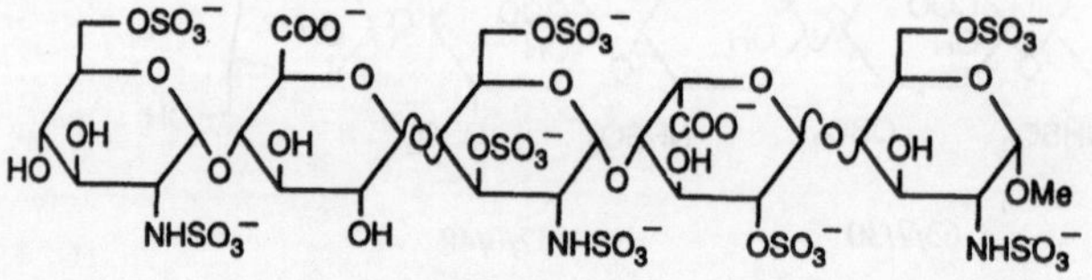

(47)   36%   64%   -

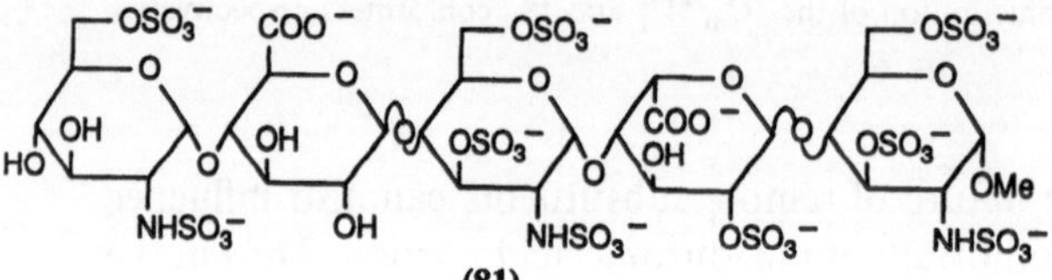

(81)   10%   90%   -

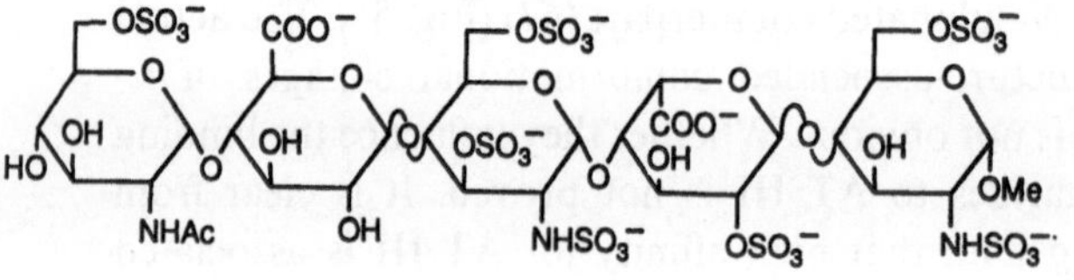

"DEFGH" (174)   10%   90%   -

Fig. 55. Conformation of iduronic acid residues in different pentasaccharides having affinity for Antithrombin III

reinforce the predominance of the gg conformer over the gt conformer, both of them being present in unsulphated glucose derivatives (*101, 107*) (Fig. 56).

Some attempts have been made to determine the conformation of synthetic heparin fragments by combining NMR and computational approaches; more precise data would be obtained from X-ray studies; however no such complex oligosaccharide has so far been obtained in crystalline form. Such an analysis has to deal with a) conformation of the iduronate ring, b) the presence of O-sulphate and N-sulphate groups for which well established stereochemical and interaction potentials are not readily available, and c) a polyanionic chain, with the associated strong polar interactions, solvent, and counter-ion effects. This problem has been addressed for the pentasaccharide (47) using molecular mechanics and ^{1}H-NMR spectroscopy (*108*). The force field used was derived from

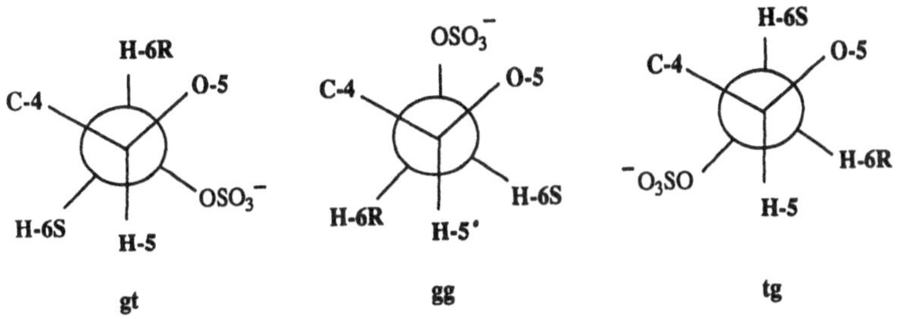

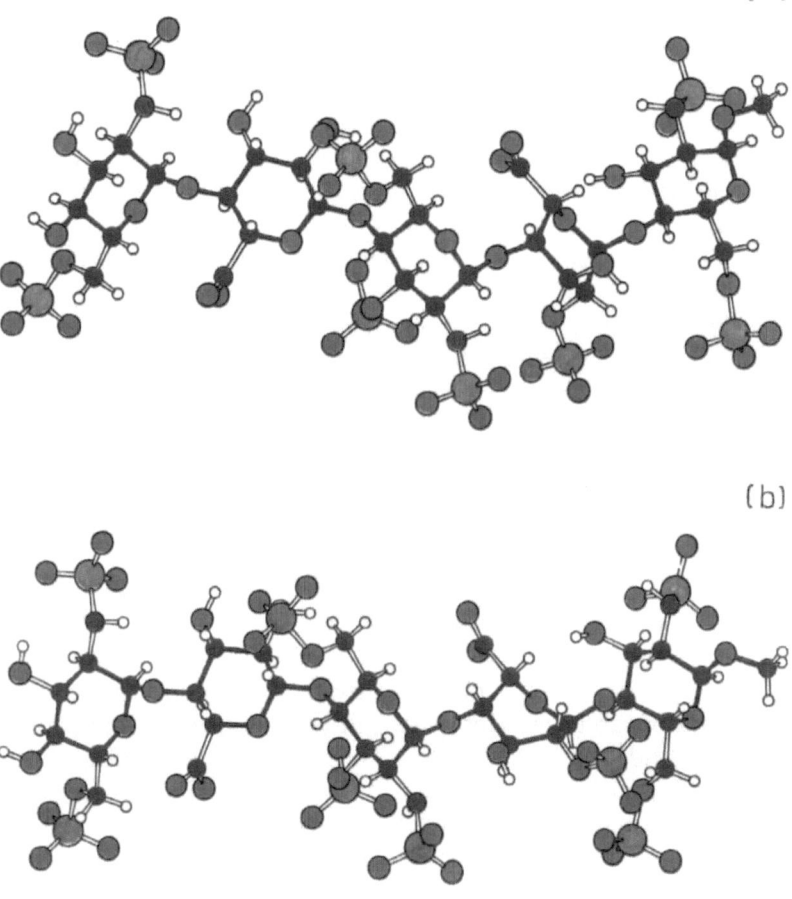

Fig. 56. Conformation about the C-5, C-6 bond of glucosamine units

(a)

(b)

Fig. 57. Representation of the two most stable conformers of pentasaccharide (47) the well known MM2 program. Concerning the uronic acid ring substituents, rotation around the C2-O2 bond of 2-O-sulphated iduronic acid shows a single energy minimum corresponding to the eclipsed arrangement of sulphur and H-2. Sulphate groups attached to the pyranose ring have been found to always adopt this conformation, with a deviation up to 30° according to the environment. In contrast free unsubstituted hydroxyl groups may assume three different conformations and the carboxylate group of uronic acids shows great freedom of rotation.

Whatever their substitution the N-sulphated glucosamine residues in (47) maintain a $^4C_1$ conformation. A single energy minimum is observed during the rotation of the C-2-N-2 bond, corresponding to the cis position with respect to H-2, as for the O-sulphate groups. The 6-sulphate has a preference for the $trans$ conformation at the C-6-O-6 bond. Among the three possible staggered conformers for the C-5, C-6 bond tg is discarded whereas gt and gg have comparable energies. NMR data are in favor of a strong predominance of the gg rotamer.

Energy maps for the different disaccharides (DE, EF, FG, GH) were plotted as a function of ϕ and ψ angles. Energy minima were selected and combined to give the most stable computed conformers of (47) in which the iduronate residue occurs either in the $^2S_0$ or the $^1C_4$ conformation (Fig. 57). From this model NOE values were computed (assuming that, according to ^{1}H-NMR 3J values, the $^2S_0$ conformer contributes up to 64% in the conformational equilibrium) and compared with the observed NOEs. Some points of disagreement between observation and computation indicated that the model could be improved. Inspection of this model shows an asymmetric distribution of essential charged groups involved in the interaction with antithrombin. This point will be a matter of discussion in the next section.

5. Interaction of Heparin Pentasaccharide Fragments with Antithrombin III

In Sects. 3 and 4 it was demonstrated that the presence and spatial orientation of particular carboxylate and sulphate groups of the heparin pentasaccharide fragment is of utmost importance to bind specifically to AT III and to induce a conformational change in AT III. The conformational change in the protease inhibitor strengthens the interaction of the heparin fragment and is required for acceleration of factor Xa neutralization.

Atha et al. (68) studied the interaction of the synthetic pentasaccharide (2) with AT III and found a binding constant of 100 nM. The synthetic tetrasaccharides consisting of residues DEFG and EFGH showed, respectively, a 100 fold and 1000 fold lower affinity for AT III as compared with the biologically active pentasaccharide (2). The non-3-O-sulphated pentasaccharide (52) was found to bind AT III only very weakly (Kd = 5.10^5 nM) and showed essentially no capacity for effecting a conformational change in AT III. These data were recently confirmed and other pentasaccharides were studied (81). VISSER et al. (80) studied the influence of the sulphate groups at unit H in the binding with AT III. The results are shown in Fig. 58. Thus, introduction of an extra 3-O-sulphate group at unit H (i.e. compound (81)) leads to a twenty fold increase of the binding affinity for AT III, whereas removal of the 6-O-sulphate group (i.e. compound (53)) from the same unit gives a two to four fold reduction of the binding affinity. Furthermore, by studying compound (100) it was found that the 3-O-sulphate at unit H (which only occurs in synthetic analogues), contributes more heavily to the AT III binding process than the 6-O-sulphate.

It may be assumed that the essential charged groups of the active pentasaccharide molecules interact with complementary lysine and arginine residues at AT III. Detailed information about the interaction of the pentasaccharide with AT III at the molecular level was obtained as the result of a study of GROOTENHUIS and van BOECKEL (109). First a molecular model of AT III was constructed on the basis of the crystal structure of α1-antitrypsin using standard homology building techniques. Then positively charged amino acid residues were highlighted which, according to other investigations, are involved in the heparin binding process. Since it was found that both the protein and the pentasaccharide fragment display an asymmetric assembly of interaction

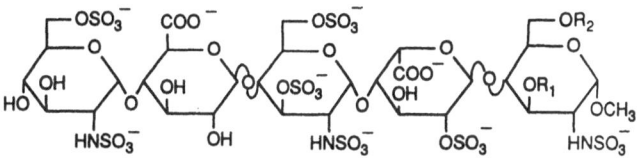

compound	R_1	R_2	Kd[nM]
(47)	H	SO_3^-	300
(53)	H	H	1200
(100)	SO_3^-	H	35
(81)	SO_3^-	SO_3^-	18

Fig. 58. The 3-O-sulphate group at the reducing end of synthetic pentasaccharides contributes more heavily in the AT III binding process than the 6-O-sulphate group

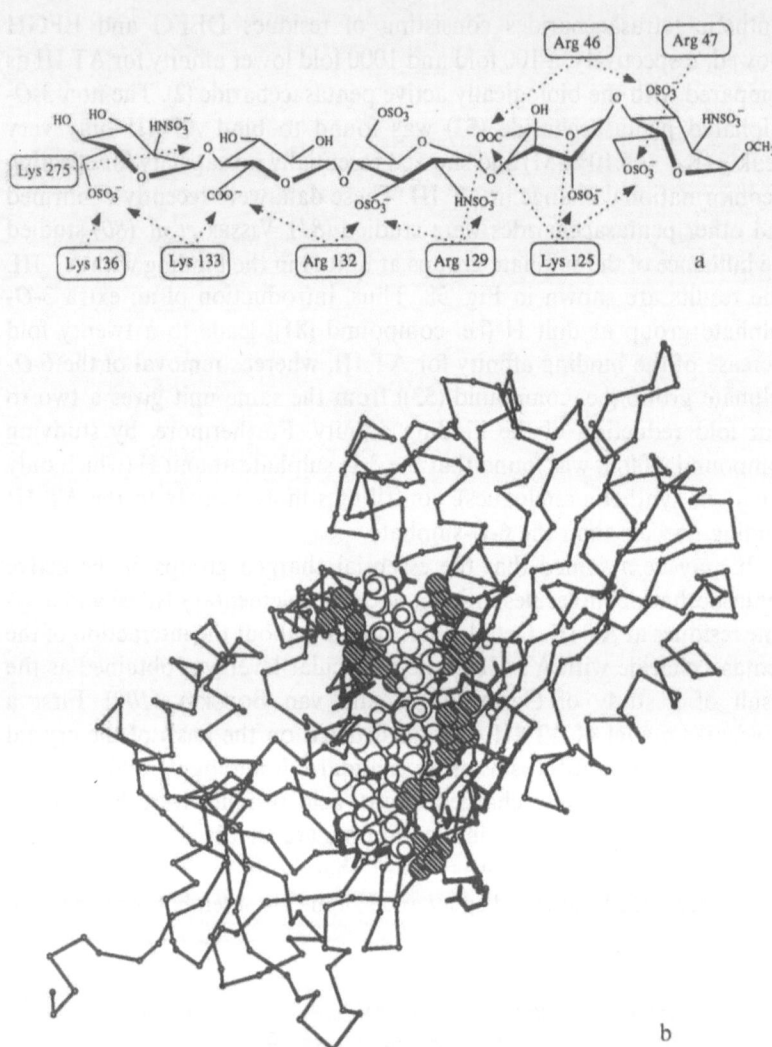

Fig. 59. a) Model of the favourable interactions between AT III and the potent pentasaccharide (8); b) Interaction of the pentasaccharide (CPK representation) with the essential amino acid residues (in black) of AT III. The solid lines represent the α-backbone of the AT III model

points, their orientation toward each other could be established during docking trials. After 5 ps of MD simulation it could be observed that residues Lys 125, Arg 129, Arg 132 and Lys 133 of AT III form an elongated binding site complementary to the negatively charged southern region of the pentasaccharide (see also Fig. 24 in Sect. 3.3) while the amino acids Arg 46 and Arg 47 interact with the northern charged groups of units G and H (Fig. 59a).

Further optimization of the model was realized by another molecular dynamics simulation in the presence of water. During the latter run Lys 136 interacts with the essential 6-O-sulphate group at unit D, while the negatively charged Asp 278 forms a hydrogen bridge with the 3-hydroxyl group of unit E. The model (Fig. 59) explains the favourable contributions of essential sulphate and carboxylate groups of the synthetic heparin fragments. For instance the observation that the introduction of an extra 3-O-sulphate group at unit E (i.e. compound (**98**) in Sect. 3.7) leads to a considerable loss of activity can be ascribed to the electrostatic repulsion between this sulphate group and Asp 278. The enhanced affinity and activity of analogues with a 3-O-sulphate at unit H can be explained by the interaction with Arg 46 in the northern binding-site. The important aminosulphate at unit H is (in the model) located in the positively charged field of a long α-helix that bears Trp 49, an amino acid which is known to undergo a conformational change during heparin binding (Fig. 59b). Although a more reliable and detailed picture of the pentasaccharide-AT III complex can only be obtained by X-ray crystallography, the model described above provided for the first time detailed insight in the highly specific binding of the heparin-pentasaccharides with AT III.

Acknowledgements

The authors wish to thank: Sjoerd van Aelst, Jan Basten, Tom Beetz, Rein van de Bosch, Jean Choay, Marcel Derrien, Theo van Dinther, Philippe Duchaussoy, Françoise Gourvenec, Peter Gootenhuis, Henk van der Heijden, Guy Jaurand, Isidore Lederman, Jean-Claude Lormeau, Hans Lucas, Dick Meuleman, Pierre Sinaÿ, Jean-Marc Strassel and Jan Vos, for their participation in the work described in this article.

References

1. MCLEAN, J.: The Thromboplastic Action of Cephalin. Am. J. Physiol. **41**, 250 (1916).

2. MCLEAN, J.: The Discovery of Heparin. Circulation **19**, 75 (1959). Eds. D.A. Lane and U. Lindhal.

3. RODEN, L.: Highlights in the History of Heparin. In "Heparin". (D.A. LANE and U. LINDAHL, eds.), pp. 1–23. London: Edward Arnold. **1989**.

4. CASU, B.: Structure and Biological Activity of Heparin. Adv. Carbohydr. Chem. Biochem. **43**, 51 (1985).

5. LINDAHL, U., D.S. FEINGOLD, and L. RODEN: Biosynthesis of Heparin. Trends Biochem. Sci. **11**, 221 (1986).
6. BJORK, I., S.T. OLSON, and J.D. SHORE: Molecular Mechanisms of the Accelerating Effect of Heparin on the Reactions between Antithrombin and Clotting Proteinases. In "Heparin". (D.A. LANE and U. LINDAHL, eds.), pp. 229–255. London: Edward Arnold. **1989**.
7. VERSTRAETE, M.: Pharmacotherapeutic Aspects of Unfractionated and Low Molecular Weight Heparins. Drugs **40**, 498 (1990).
8. JACQUES, L.B.: Heparins-Anionic Polyelectrolyte Drugs. Pharmacol. Rev. **31**, 99 (1979).
9. CLOWES, A.W. and M.J. KARNOVSKY: Suppression by Heparin of Smooth Muscle Cell Proliferation in Injured Arteries. Nature **265**, 625 (1977).
10. FOLKMAN, J.: Regulation of Angiogenesis: A New Function of Heparin. Biochem. Pharmacol. **34**, 905 (1985).
11. BABA, M., R. PAUWELS, J. BALZARINI, J. DESMYTER, and E. De CLERCQ: Antiviral Activity of Heparin and Dextran Sulphate against Human Immunodeficiency Virus (HIV) in Vitro. Ann. N.Y. Acad. Sci. **556**, 419 (1989).
12. ANDERSSON, L.O., T.W. BARROWCLIFFE, E. HOLMER, E.A. JOHNSON, and G.E.C. SIMS: Anticoagulant Properties of Heparin Fractionated by Affinity Chromatography on matrix-bound Antithrombin III and by Gel Filtration. Thromb. Res. **9**, 575 (1976).
13. HOOK, M., I. BJORK, J. HOPWOOD, and U. LINDAHL: Anticoagulant Activity of Heparin: Separation of High Activity and Low Activity Heparin Species by Affinity Chromatography on Immobilized Antithrombin. FEBS Lett. **66**, 90 (1976).
14. LAM, L.H., J.E. SILBERT, and R.D. ROSENBERG: The Separation of Active and Inactive Forms of Heparin. Biochem. Biophys. Res. Commun. **69**, 570 (1976).
15. CHOAY, J., J.C. LORMEAU, M. PETITOU, P. SINAY, and J. FAREED: Structural Studies on a Biologically Active Hexasaccharide Obtained from Heparin. Ann. N.Y. Acad. Sci. **370**, 644 (1981).
16. THUNBERG, L., G. BACKSTROM, and U. LINDAHL: Further Characterization of the Antithrombin-binding Sequence in Heparin. Carbohydr. Res. **100**, 393 (1982).
17. LINDAHL, U., G. BACKSTROM, L. THUNBERG, and I.G. LEDER: Evidence for a 3-*O*-sulphated D-glucosamine Residue in the Antithrombin-binding Sequence of Heparin. Proc. Natl. Acad. Sci. USA **77**, 6551 (1980).
18. KLEMER, A., and U. KRASKA: Synthese von Athyl-2-Amino-2-Desoxy-4-*O*-(β-D-Glucuronopyranosyl)-α, β-D-Glucopyranosid. Tetrahedron Lett. **13**, 431 (1972).
19. KISS, J., and P. TASCHNER: Synthesis of Heparin Saccharides. VI. Synthesis and Reactivity of some 4-*O*-(α-D-Hexopyranosyl)-α-D-Glucopyranosiduronate Derivatives. J. Carbohydr. Nucl. Nuc. **4**, 101 (1977).
20. KISS, J., and P.C. WYSS: Synthesis of Heparin Saccharides. Stereospecific Synthesis of Derivatives of 2-Amino-2-Deoxy-4-O-(α-D-Glucopyranuronosyl)-D-Glucose. Tetrahedron Lett. **13**, 3055 (1972).
21. KISS, J., and P.C. WYSS: Synthesis of Heparin Saccharides. II. Synthesis and Stereochemical Aspects of Anomeric Methyl (Benzyl 2,3-di-*O*-Benzyl-L-Idopyranosid) Uronates. Carbohydr. Res. **27**, 282 (1973).
22. KISS, J., and P.C. WYSS: Synthesis of Heparin Saccharides. V. Anomeric *O*-Benzyl Derivatives of L-Idopyranosyluronic Acid. Tetrahedron **32**, 1399 (1976).
23. WYSS, P.C., and J. KISS: Synthesis of Heparin Saccharides. III. Synthesis of Derivatives of D-Glucosamine as Starting Materials for Disaccharides. Helv. Chim. Acta **58**, 1833 (1975).
24. WYSS, P.C., J. KISS, and W. ARNOLD: Synthesis of Heparin Saccharides. IV. Synthesis

of Disaccharides Possessing the Structure of a Repeating Unit of Heparin. Helv. Chim. Acta **58**, 1847 (1975).

25. PAULSEN, H.: Advances in Selective Chemical Syntheses of Complex Oligosaccharides. Angew. Chem. Int. Ed. Engl. **21**, 155 (1982).

26. SCHMIDT, R.R.: New Methods for the Synthesis of Glycosides and Oligosaccharides-Are There Alternatives to the Koenigs-Knorr Method? Angew. Chem. Int. Ed. Engl. **25**, 212 (1986).

27. WESSEL, H.P., Alkylating γ-Lactone-Opening: a short Synthesis of benzyl 3-*O*-Benzyl-1,2-*O*-Isopropylidene-α-D-Glucofuranuronate. J. Carbohydr. Chem. **8**, 443–455 (1989).

28. SINAY, P., J.C. JACQUINET, M. PETITOU, P. DUCHAUSSOY, I. LEDERMAN, J. CHOAY, and G. TORRI: Total Synthesis of a Heparin Pentasaccharide Fragment having High Affinity for Antithrombin III. Carbohydr. Res. **132**, C5 (1984).

29. PETITOU, M., P. DUCHAUSSOY, I. LEDERMAN, J. CHOAY, P. SINAY, J.C. JACQUINET, and G. TORRI: Synthesis of Heparin Fragments. A Chemical Synthesis of the Pentasaccharide   *O*-(2-Deoxy-2-Sulfamido-6-*O*-Sulfo-α-D-Glucopyranosyl)-(1 → 4)-*O*-(β-D-Glucopyranosyluronic   Acid)-(1 → 4)-*O*-(2-Deoxy-2-Sulfamido-3,6-di-*O*-Sulfo-α-D-Glucopyranosyl)-(1 → 4)-*O*-(2-*O*-Sulfo-α-L-Idopyranosyluronic   Acid)-(1 → 4)-2-Deoxy-2-Sulfamido-6-*O*-sulfo-D-Glucopyranose Decasodium Salt, a Heparin Fragment Having High Affinity for Antithrombin III. Carbohydr. Res. **147**, 221 (1986).

30. ZISSIS, E., and H.G. FLETCHER Jr.: Benzyl 2,3,4-Tri-*O*-Benzyl-β-D-Glucopyranosiduronic Acid and some Related Compounds. Carbohydr. Res. **12**, 361 (1970).

31. MEHLTRETTER, C.L.: D-Glucuronic acid: α-D-Glucofuranurono-6,3-Lactone by Catalytic Air Oxidation of 1,2-*O*-Isopropylidene-α-D-Glucofuranose. Meth. Carbohydr. Chem. vol II, 29 (1963).

32. NAKAHARA, Y., and T. OGAWA: Synthesis of Methyl (Allyl 2,3-di-*O*-Benzyl-β-D-Galactopyranosid)Uronate and Methyl (2,3-di-*O*-Benzyl-α- and β-D-Galactopyranosyl Fluoride)Uronate. Carbohydr. Res. **173**, 306 (1988).

33. NAKAHARA, Y., and T. OGAWA: Stereoselective Total Synthesis of Dodecagalacturonic Acid, a Phytoalexin Elicitor of Soybean. Carbohydr. Res. **205**, 147 (1990).

34. VAN BOECKEL, C.A.A., T. BEETZ, J.N. VOS, A.J.M. de JONG, S.F. van AELST, R.H. van den BOSCH, J.M.R. MERTENS, and van der VLUGT, F.A.: Synthesis of a Pentasaccharide Corresponding to the Antithrombin III Binding Fragment of Heparin. J. Carbohydr. Chem. **4**, 293 (1985).

35. MEYER, A.S., and T. REICHSTEIN: L-Idose aus D-Glucose, sowie ein neuer Weg zur L-Idomethylose. Helv. Chem. Acta **29**, 152 (1946).

36. PERCHEMLIDES, P., T. OSAWA, E.A. DAVIDSON, and R.W. JEANLOZ: Synthesis of α-L-Idopyranosyl, (α-L-Idopyranosyluronic Acid), α-D-Mannopyranosyl, and (α-D-Mannopyranosyluronic Acid) Phosphates. Carbohydr. Res. **3**, 463 (1967).

37. DAX, K., I. MACHER, and H. WEIDMANN: Reaktionen der D-glucuronsäure. 8.Mitt. Synthese von Derivaten der L-Idofuranose und des D-Mannofuranurono-6,3-Lactons aus D-Glucofuranurono-6,3-Lacton. J. Carbohydr. Nuc. Nuc. **1**, 323 (1974).

38. BLANC-MUESSER, M., J. DEFAYE, D. HORTON, and J.H. TSAI: L-Idose and L-Iduronic Acid. Meth. Carbohydr. Chem. **8**, 177 (1980).

39. BAGGET, N., and A.K. SAMRA: Re-Examination of the Acid Hydrolysis of 5,6-Anhydro-1,2-*O*-Isopropylidene-β-L-Idofuranose. Carbohydr. Res. **127**, 149 (1984).

40. LEHMANN, J.: Reaktionen Enolischer Zuckerderivate. Teil 1. Hydroborierung enolischer Zuckerderivate, ein Weg zur Darstellung schwer zugänglicher Hexosen und zur spezifischen Markierung mit Tritium. Carbohydr. Res. **2**, 1 (1966).

41. NASSR, M.A.M., M. PETITOU, J. CHOAY, and P. SINAY: Synthèse de Disaccharides

Contenant le L-Idopyrannose à l'Extrémité non-Réductrice. Xèmes Journées sur la Chimie et la Biochimie des Glucides, Paris, 5–7 juillet 1982.

42. ICHIKAWA, Y., and H. KUZUHARA: Synthesis of 1,6-Anhydro-2,3-di-O-Benzoyl-4-O-(methyl-2,3,4-tri-O-Benzoyl-α-L-Idopyranosyluronate)-β-D-Glucopyranose from Cellobiose. Carbohydr. Res 115, 117 (1983).

43. BAGGET, N., and A. SMITHSON: Synthesis of L-Iduronic Acid Derivatives by Epimerisation of Anancomeric D-Glucuronic Acid Analogues. Carbohydr. Res. 108, 59 (1982).

44. CHIBA, T., and P. SINAY: Application of a Radical Reaction to the Synthesis of L-Iduronic acid Derivatives from D-Glucuronic Acid Analogues. Carbohydr. Res. 151, 379 (1986).

45. CHIDA, N., E. YAMADA, and S. OGAWA: Synthesis of Methyl (Methyl D- and L-Idopyranosid)uronates from Myo-Inositol. J. Carbohydr. Chem. 7, 555 (1988).

46. ICHIKAWA, Y., R. MONDEN, and H. KUZUHARA: Synthesis of a Heparin Pentasaccharide Fragment with a High Affinity for Antithrombin III Employing Cellobiose as a Key Starting Material. Tetrahedron Lett. 27, 611 (1986).

47. ICHIKAWA, Y., A. MANAKA, and H. KUZUHARA: Discrimination between the 2,3- and the 2',3'-Hydroxyl Groups of Maltose and Cellobiose through their Specific Protection. Carbohydr. Res. 138, 55 (1985).

48. ICHIKAWA, Y., R. ICHIKAWA, and H. KUZUHARA: Synthesis from Cellobiose, of a Trisaccharide Closely Related to the GlcNAc → GlcA → GlcN Segment of the Antithrombin-binding Sequence of Heparin. Carbohydr. Res. 141, 273 (1985).

49. ICHIKAWA, Y., R. MONDEN, and H. KUZUHARA: Synthesis of Methyl Glycoside Derivatives of Tri- and Penta-saccharides Related to the Antithrombin III-binding Sequence of Heparin, employing Cellobiose as a Key Starting Material. Carbohydr. Res. 172, 37 (1988).

50. SHING, T.K.M., and A.S. PERLIN: Synthesis of Benzyl 2-Azido-2-Deoxy-4-O-β-D-Glucopyranosyl-α-D-Glucopyranoside and 1,6-Anhydro-2-Azido-2-Deoxy-4-O-β-D-Glucopyranosyl-β-D-Glucopyranose. Carbohydr. Res. 130, 65 (1984).

51. GLUSHKA, J.N., D.N. GUPTA, and A.S. PERLIN: The Conversion of Maltose into Disaccharides having 2-Amino-2-Deoxy-α-D-Glucose and L-Idose as Constituent Sugars, for the Synthesis of Model Compounds Related to Heparin. Carbohydr. Res. 124, C12 (1983).

52. GLUSHKA, J.N., and A.S. PERLIN: Formation of Disaccharides related to Heparin and Heparan Sulphate by Chemical Modification of Maltose. Carbohydr. Res. 205, 305 (1990).

53. UENO, Y., K. HORI, R. YAMAUCHI, M. KISO, A. HASEGAWA, and K. KATO: Reaction of Maltose with 2,2-Dimethoxypropane. Carbohydr. Res. 89, 271 (1981).

54. PETITOU, M., P. DUCHAUSSOY, I. LEDERMAN, J. CHOAY, J.C. JACQUINET, P. SINAY, and G. TORRI: Synthesis of Heparin Fragments: A Methyl α-Pentaoside with High Affinity for Antithrombin III. Carbohydr. Res. 167, 67 (1987).

55. PETITOU, M., G. JAURAND, M. DERRIEN, P. DUCHAUSSOY, and J. CHOAY: A New Highly Potent, Heparin-like Pentasaccharide Fragment Containing a Glucose Residue instead of a Glucosamine. BioMed. Chem. Lett. 1, 95 (1991).

56. WALENGA, J.M., J. FAREED, M. PETITOU, M. SAMAMA, J.C. LORMEAU, and J. CHOAY: Intravenous Antithrombotic Activity of a Synthetic Heparin Pentasaccharide in a Human Serum Induced Stasis Thrombosis Model. Thromb. Res. 43, 243 (1986).

57. WALENGA, J.M., M. PETITOU, J.C. LORMEAU, M. SAMAMA, J. FAREED, and J. CHOAY: Antithrombotic Activity of a Synthetic Heparin Pentasaccharide in a Rabbit Stasis

Thrombosis Model using Different Thrombogenic Challenges. Thromb. Res. **46**, 187 (1987).

58. HOBBELEN, P.M.J., T.G. van DINTHER, G.M.T. VOGEL, C.A.A. van BOECKEL, H.C.T. MOELKER, D.G. MEULEMAN: Pharmacological Profile of the Chemically Synthesized Antithrombin III Binding Fragment of Heparin (pentasaccharide) in Rats. Thromb. Haemost. **63**, 265–270 (1990).

59. MEULEMAN, D.G., P.M.J. HOBBELEN, T.G. van DINTHER, G.M.T. VOGEL, C.A.A. van BOECKEL, and H.C.T. MOELKER: Anti-factor Xa Activity and Antithrombotic Activity in Rats of Structural Analogues of the Minimum Antithrombin III binding Sequence: Discovery of Compounds with A Longer Duration of Action than of the Natural Pentasaccharide. Semin. Thromb. Hemostasis **17**, 112 (1991).

60. LOGANATHAN, D., H.M. WANG, L.M. MALLIS, and R.J. LINHARDT: Structural Variation in the Antithrombin III Binding Site Region and its Occurrence in Heparin from Different Sources. Biochemistry **29**, 4362 (1990).

61. DUCHAUSSOY, P., P.S. LEI, M. PETITOU, P. SINAY, J.C. LORMEAU, and J. CHOAY: The First Total Synthesis of the Antithrombin III Binding Site of Porcine Mucosa Heparin. BioMed. Chem. Lett **1**, 99 (1991).

62. LINDAHL, U., G. BACKSTROM, and L. THUNBERG: The Antithrombin-Binding Sequence in Heparin. Identification of an essential 6-O-Sulfate Group. J. Biol. Chem. **258**, 9826 (1983).

63. ATHA, D.H., J.C. LORMEAU, M. PETITOU, R.D. ROSENBERG, and J. CHOAY: Contribution of Monosaccharide Residues in Heparin Binding to Antithrombin III. Biochemistry **24**, 6723 (1985).

64. RIESENFELD, J., L. THUNBERG, M. HOOK, and U. LINDAHL: The Antithrombin-Binding Sequence of Heparin. Location of Essential N-Sulfate Groups. J. Biol. Chem. **256**, 2389 (1981).

65. PETITOU, M.: Synthetic Heparin Fragments: New and Efficient Tools for the Study of Heparin and its Interactions. Nouv. Rev. Fr. Hematol. **26**, 221 (1984).

66. CHOAY, J.: Biologic Studies on Chemically Synthesized Pentasaccharide and Tetrasaccharide Fragments. Semin. Thromb. Hemostasis **11**, 81 (1985).

67. PETITOU, M., P. DUCHAUSSOY, L. LEDERMAN, J. CHOAY, and P. SINAY: Binding of Heparin to Atithrombin III: a Chemical Proof of the Critical Role played by a 3-Sulfated-2-Amino-2-Deoxy-D-Glucose Residue. Carbohydr. Res. **179**, 163 (1988).

68. ATHA, D.H., J.-C. LORMEAU, M. PETITOU, R.D. ROSENBERG, and J. CHOAY: Contribution of 3-O- and 6-O-Sulfated Glucosamine Residues in the Heparin Induced Conformational Change in Antithrombin III, Biochemistry **26**, 6454 (1987).

69. BEETZ, T., and C.A.A. van BOECKEL: Synthesis of an Antithrombin Binding Heparin-like Pentasaccharide lacking 6-O-Sulfate at its Reducing End. Tetrahedron Lett. **27**, 5889 (1986).

70. PETITOU, M., J.C. LORMEAU, and J. CHOAY: Interaction of Heparin and Antithrombin III. The Role of O-Sulfate Groups. Eur. J. Biochem. **88**, 637 (1988).

71. PETITOU, M., P. DUCHAUSSOY, and J. CHOAY: p-Anisyl Ethers in Carbohydrate Chemistry: Selective Protection of the Primary Alcohol Function. Tetrahedron Lett. 1389, (1988).

72. van BOECKEL, C.A.A., et al.: unpublished results.

73. AGARWAL, A., and I. DANISHEFSKY: Requirement of free Carboxyl Groups for the Anticoagulant Activity of Heparin, Thromb. Res. **42**, 673 (1986).

74. van BOECKEL, C.A.A., H. LUCAS, S.F. van AELST, M.W.P. van den NIEUWENHOF, G.N. WAGENAARS, and J.-R. MELLEMA: Synthesis and Conformational Analysis of an

Analogue of the Antithrombin-binding Region of Heparin: the Role of the Carboxylate Function of α-L-Idopyranuronate. Recl. Trav. Chim. Pays-Bas **106**, 581 (1987).

75. van Aelst, S.F., and C.A.A. van Boeckel: Synthesis of an Analogue of the Antithrombin Binding Region of Heparin containing α-L-Idopyranose; Recl. Trav. Chim. Pays-Bas **106**, 593 (1987).

76. Vos, J., *et al.*: unpublished results.

77. Petitou, M., *et al.*: unpublished results.

78. van Boeckel, C.A.A., T., Beetz, and S.F. van Aelst: Synthesis of a potent Antithrombin activating Pentasaccharide: A new Heparin-like Fragment Containing two 3-O-Sulphated Glucosamines. Tetrahedron Lett. 803 (1988).

79. van Boeckel, C.A.A., S.F. van Aelst, T. Beetz, D.G. Meuleman, Th.G. van Dinther, and H.C.T. Moelker: Structure-Activity Relationships of Synthetic Heparin Fragments: Discovery of a very Potent AT-III Activating Pentasaccharide. Ann. N.Y. Acad. Sci. **556**, 489, 1989.

80. Visser, A., M.T. Buiting, T.G. van Dinther, C.A.A. van Boeckel, P.D.J. Grootenhuis, and D.G. Meuleman: The AT-III Binding Affinities of a Series of Synthetic Pentasaccharide Analogues. Thromb. Haemost. **65**, 1296 (1991).

81. Barzu, T., M. Petitou, G. Jaurand, J.C. Lormeau, and J. Choay: Binding to Antithrombin III of the synthetic Oligosaccharides derived from the High Affinity Pentasaccharide Sequence of Heparin. Thromb. Haemost. **65**, 934 (1991).

82. Petitou, M., J.C. Lormeau, and J. Choay: A New Synthetic Pentasaccharide with Increased Anti-Factor Xa Activity: Possible Role for Anionic Clusters in the Interaction of Heparin and Antithrombin III. Semin. Thromb. Hemostasis **17**, 143, (1991).

83. Basten, J., G. Jaurand, *et al.*: unpublished results.

84. Basten, J., *et al.*: unpublished results.

85. Kat-Vanden Nieuwenhof, M.W.P., J.E.M. Basten, M. Lucas, and C.A.A. van Boeckel: Synthesis of some very potent Antithrombin III activating Heparin-Like Fragments. Fifth European Symposium on Carbohydrates, Eurocarb V, Prague, 21–25 August 1989. Abstr. A-39.

86. Petitou, M., G. Jaurand, M. Derrien, P. Duchaussoy, and J. Choay: Synthesis of selectively oversulfated Heparin-Like Pentasaccharides with high anti-factor Xa Activity. Fifth European Symposium on Carbohydrates, Eurocarb V, Prague, 21–25 August 1989. Abstr. A-68.

87. van Boeckel, C.A.A., G.N. Wagenaars, and J.R. Mellema: Conformational Analysis of a Biological Active Heparin-like Compound, which Contains an Open Chain Fragment. Recl. Trav. Chim. Pays-Bas **107**, 649 (1988).

88. Lucas, H., J.E.M. Basten, Th.G. van Dinther, D.G., Meuleman, S.F. van Aelst, and C.A.A. van Boeckel: Synthesis of Heparin-Like Pentamers Containing "Opened" Uronic Acid Moieties. Tetrahedron **46**, 8207 (1990).

89. Wessel, H.P., L. Labler, and T.B. Tschopp: Synthesis of an N-Acetylated Heparin Pentasaccharide and its Anticoagulant Activity in Comparison with the Heparin Pentasaccharide with High anti-Factor-Xa Activity. Helv. Chem. Acta **72**, 1268 (1989).

90. Kraaijeveld, N.A., and C.A.A. van Boeckel: Synthesis of Several Sulphated and Non-Sulphated Pentasaccharides, corresponding to the *E. Coli* K5 Glycosaminoglycan. Recl. Trav. Chim. Pays-Bas **108**, 39 (1989).

91. Vos, J.N., P. Westerduin, and C.A.A. van Boeckel: Synthesis of a 6-O-Phosphorylated Analogue of the Antithrombin III Binding Sequence of Heparin. BioMed. Chem. Lett. **1**, 143 (1991).

92. KANYO, Z.F., and D.W. CHRISTIANSON: Biological Recognition of Phosphate and Sulfate. J. Biol. Chem. **266**, 4264 (1991).

93. EDGE, A.S.B., and R.G. SPIRO: Characterization of novel Sequences Containing 3-O-Sulfated Glucosamine in Glomerular Basement Membrane Heparan Sulfate and Localization of Sulfated Disaccharides to a Peripheral Domain. J. Biol. Chem. **265**, 15874 (1990).

94. NUKADA, T., H. LUCAS, P. KONRADSSON, and C.A.A. van BOECKEL,: Syntheses of larger Modified Oligosaccharides Containing "Opened Carbohydrate" Fragments. Synlett **365** (1991).

95. LEMIEUX, R.U., K.B. HENDRIKS, R.V. STICK, and K. JAMES: Halide Ion Catalyzed Glycosidation Reactions. Synthesis of α-linked Disaccharides. J. Am. Chem. Soc. **97**, 4056 (1975).

96. LUCAS, H., J. BASTEN, P. KONRADSSON, B. OLDE HANTER, C.A.A. van BOECKEL, G, JAURAND, P. DUCHAUSSOY, M. DERRIEN, and M. PETITOU: Syntheses and Structure-Activity Relationships of some new Potent Analogues of Heparin; Preparation of Alkylated "Non-Glycsoaminoglycans". Presented at Eurocarb VI, VIth European Symposium Carbohydrate Chemistry, Edinburgh Sept. 1991. Abstract B. 170.

97. MEULEMAN, D., et al.: to be published.

98. GATTI, G., B. CASU, G.K. HAMER, and A.S. PERLIN: Studies on the Conformation of Heparin by ^{1}H- and ^{13}C-NMR Spectroscopy. Macromolecules **12**, 1001 (1979).

99. TORRI, G., B. CASU, G. GATTI, M. PETITOU, J. CHOAY, J.C. JACQUINET, and P. SINAY: Mono- and Bidimensional 500 MHz ^{1}H-NMR Spectra of a Synthetic Pentasaccharide Corresponding to the Binding Sequence of Heparin to Antithrombin III: Evidence for Conformational Peculiarity of the Sulphated Iduronate Residue. Biochem. Biophys. Res. Commun. **128**, 134 (1985).

100. RAGAZZI, M., D.R. FERRO, and A. PROVASOLI: A Force-field Study of the Conformational Characteristics of the Iduronate Ring. J. Comput. Chem. **7**, 105 (1986).

101. van BOECKEL, C.A.A., S.F. van AELST, G.N. WAGENAARS, J.R. MELLEMA, H. PAULSEN, J. PETERS, A. POLLEX, and V. SINNWELL: Conformational Analysis of Synthetic Heparin-like Oligosaccharides Containing α-L-Idopyranosyluronic Acid. Recl. Trav. Chim. Pays-Bas **106**, 19 (1987).

102. FERRO, D.R., A. PROVASOLI, M. RAGAZZI, B. CASU, G. GATTI, G. TORRI, V. BOSSENNEC, B. PERLY, P. SINAY, M. PETITOU, and J. CHOAY: Conformer Populations of L-Iduronate Acid Residue in Glycosaminoglycan Sequences. Carbohydr. Res. **195**, 157 (1990)

103. FERRO, D.R., A. PROVASOLI, M. RAGAZZI, B. CASU, G. GATTI, J.C. JACQUINET, P. SINAY, M. PETITOU, and J. CHOAY: Evidence for Conformational Equilibrium of the Sulphated L-Iduronate Residue in Heparin and in Synthetic Heparin Mono- and Oligosaccharides: MNR and Force-field Studies. J. Am. Chem. Soc. **108**, 6773 (1986).

104. SANDERSON, P.N., T.N. HUCKERBY, and I.A. NIEDUSZYNSKI: Conformational Equilibrium of Unsulphated Iduronate in Heparan Sulphate Tetrasaccharides. Glycoconjugate J. **2**, 109 (1985).

105. PAULSEN, H., A. POLLEX, V. SINNWELL, and C.A.A. van BOECKEL: Konformationsanalyse von Heparin-analogen Di- und Trisacchariden mit α-L-Idopyranose-Einheiten. Liebigs Ann. Chem. 411 (1988).

106. MEYER, B., and R. STUIKE-PRILL: Syntheses of Benzyl 6-O-Sulfo-β-D-Glucopyranoside Salts and their 6-S-Deuterated Analogues. Conformational Preferences of their (Sulfonyloxy)methyl Group. J. Org. Chem. **55**, 902 (1990).

107. NISHIDA, Y., H. HORI, H. OHRUI, and H. MEGURO: ^{1}H NMR Analyses of Rotameric

Distribution of C5-C6 Bonds of Glucopyranoses in Solution. J. Carbohydr. Chem. 7, 239 (1988).

108. RAGAZZI, M., D.R. FERRO, B. PERLY, P. SINAY, M. PETITOU, and J. CHOAY: Conformation of the Pentasaccharide Corresponding to the Binding Site of Heparin for Antithrombin III. Carbohydr. Res. 195, 169 (1990).

109. GROOTENHUIS, P.D.J., and C.A.A. van BOECKEL: Constructing a Molecular Model of the Interaction between Antithrombin III and a Potent Heparin Analogue. J. Am. Chem. Soc. 113, 2743 (1991).

110. JAURAND, G., J. BASTEN, I. LEDERMAN, C.A.A. van BOECKEL, M. PETITOU: Biologically Active Heparin-Like Fragments with a "Non-Glycosamino" glycan Structure. Part 1: A Pentasaccharide Containing a 3-O-Methyl Iduronic Acid Unit. BioMed. Chem. Lett. 2, 897 (1992).

111. BASTEN, J., G. JAURAND, B. OLDE-HANTER, M. PETITOU, C.A.A. van BOECKEL: Biologically Active Heparin-like Fragments with a "Non-Glycosamino"glycan Structure. Part 2: A Tetra-O-Methylated Pentasaccharide with High Affinity for Antithrombin III. BioMed. Chem. Lett. 2, 901 (1992).

112. BASTEN, J., G. JAURAND, B. OLDE-HANTER, P. DUCHAUSSOY, M. PETITOU, C.A.A. van BOECKEL: Biologically Active Heparin-like Fragments with a "Non-Glycosamino"glycan Structure. Part 3: O-Alkylated-O-Sulphated Pentasaccharides. BioMed. Chem. Lett. 2, 905 (1992).

(Received March 5, 1992)

Author Index

Subject Index

General Index
Vols. 21–60

Author Index

Vols. 21–60

Subject Index

Vols. 21–60

Fortschritte der Chemie organischer Naturstoffe

Progress in the Chemistry of Organic Natural Products

Founded by L. Zechmeister.
Edited by W. Herz, G.W. Kirby, R.E. Moore, W. Steglich, and C. Tamm

Volume 59:
1992. 1 figure. IX, 328 pages. Cloth DM 260,–, öS 1820,–.
ISBN 3-211-82278-X
Contents: Shin-Ichi Hatanaka: Amino Acids from Mushrooms. • I. Wahlberg and
A.-M. Eklund: Cembranoids, Pseudopteranoids, and Cubitanoids of Natural
Occurrence.

Volume 58:
1991. 64 figures. VII, 343 pages. Cloth DM 280,–, öS 1960,–.
ISBN 3-211-82265-8
Contents: J.A. Robinson: Chemical and Biochemical Aspects of Polyether-
Ionophore Antibiotic Biosynthesis. • R.D.H. Murray: Naturally Occurring Plant
Coumarins.

Volume 57:
1991. 26 figures and 2 plates. X, 212 pages. DM 210,–, öS 1470,–.
ISBN 3-211-82245-3
Contents: P. Metzger, C. Largeau, E. Casadevall: Lipids and Macromolecular
Lipids of the Hydrocarbon-rich Microalga *Botryococcus braunii*. Chemical
Structure and Biosynthesis. Geochemical and Biotechnological Importance. •
D.P. Chakraborty and S. Roy: Carbazole Alkaloids III. • G. R. Pettit: The
Bryostatins.

Springer-Verlag Wien New York

Volume 56:
1991. 8 figures. X, 188 pages. Cloth DM 220,–, öS 1540,–.
ISBN 3-211-82188-0
Contents: J. Asselineau: Bacterial Lipids Containing Amino Acids or Peptides Linked by Amine Bonds. • J.Kagan: Naturally Occurring Di- and Trithiophenes.

Volume 55:
1989. 41 figures. X, 208 pages. Cloth DM 190,–, öS 1330,–.
ISBN 3-211-82087-6
Contents: M.T. Davies-Coleman and D.E.A. Rivett: Naturally Occurring 6-substitutes 5,6-dihydro-α-pyrones • K. Krohn: Building Blocks for the Total Synthesis of Anthracyclinones • M. Lounasmaa and J. Galambos: Indole Alkaloid Production in Catharanthus Roseus Cell Suspension Cultures • C.E. James, L. Hough, and R. Khan: Sucrose and Its Derivatives.

Volume 54:
1988. VII, 353 pages. Cloth DM 320,–, öS 2240,–.
ISBN 3-211-82086-8
Contents: T. Murakami and N. Tanaka: Occurrence, Structure and Taxonomic Implications of Fern Constituents.

Volume 53:
1988. 72 figures. VIII, 311 pages. Cloth DM 275,–, öS 1930,–.
ISBN 3-211-82074-4
Contents: L.F. Alves: Chemical Ecology and the social Behavior of Animals • T. Nomura: Phenolic Compounds of the Mulberry Tree and Related Plants • A. Chimiak and M.J. Milewska: N-Hydroxyamino Acids and Their Derivatives.

All Volumes and Cumulative Index 1-20 available
Price reduction for subscribers: 10%

Special reduced price (20% reduction) for the complete Series Vols. 1-60
incl. the Cumulative Index to Vols. 1 20

Springer-Verlag Wien New York